T0271098

EU Trade-Related Measures against Illegal Fishing

Focusing on the experiences of Thailand and Australia, this book examines the impact of trade-restrictive measures as related to the EU's regulations to prevent Illegal, Unreported and Unregulated (IUU) fishing.

It is widely accepted that IUU fishing is harmful, and should be stopped, but there are different approaches to tackling it. Acknowledging this, this book argues that major efforts to fight IUU fishing require careful analyses if the goal is to achieve optimal results and avoid unintended consequences. The book draws on the recent experiences of Thailand and Australia to offer an empirical examination of one increasingly prominent solution, trade-restrictive measures. With Thailand representing direct, active intervention by the EU and Australia a more indirect dispersion of policy narratives and discourses, the book provides a rounded view on how likely it is that different countries in different situations will adapt to the changing policy norms regarding IUU fishing. Understanding the reactions of countries who might be targeted, or otherwise be influenced by the policy, generates new knowledge that helps inform a more effective and scalable implementation of the policy on the part of the EU and a better governance preparedness on the part of non-EU fishing nations. In broader terms, this book exposes a key moment of change in the compatibility between environmental regulations and international trade. The EU IUU policy is a prime example of a policy that uses the mechanisms of international trade to account for environmental and conservation objectives. By way of the unilateral and trade-restrictive stance against IUU fishing, the EU has positioned itself as a major market and normative power, driving its sustainability norms outwards. This book sheds light on the efficacy of this policy setup based on the analysis of country perspectives, which is a key factor influencing its potential spread.

This book will be of interest to students and scholars of international fisheries politics, marine conservation, environmental policy and international trade, and will also be of interest to policymakers working in these areas.

Alin Kadfak is a Researcher at the Department of Rural and Urban Development, Swedish University of Agricultural Sciences, Uppsala, Sweden. Her research interests are resource governance, migration, ethical food systems, policy discourses, supply chains and development in the fisheries sector.

Kate Barclay is a Professor and Director of the Climate, Society and Environment Research Centre in the Faculty of Arts and Social Sciences at University of Technology Sydney, Australia. Her research interests are the governance of marine areas and resources, including social and economic seafood value chains, social inclusion in fisheries, and the well-being of people in coastal communities.

Andrew M. Song was an ARC Discovery Early Career Research Fellow (DECRA) and a member of the Climate, Society and Environment Research Centre (C-SERC) at the University of Technology, Sydney, Australia. After completing a PhD at Memorial University of Newfoundland in 2014, he spent three years at the ARC Centre of Excellence for Coral Reef Studies in James Cook University in 2016 working as a research fellow. During this time, he also held a joint affiliation with WorldFish.

Routledge Focus on Environment and Sustainability

For more information about this series, please visit: www.routledge.com/
Routledge-Focus-on-Environment-and-Sustainability/book-series/
RFES

EU Trade-Related Measures against Illegal Fishing

Policy Diffusion and Effectiveness in Thailand and Australia

Alin Kadfak, Kate Barclay, and Andrew M. Song

LONDON AND NEW YORK

First published 2023
by Routledge
4 Park Square, Milton Park, Abingdon, Oxon OX14 4RN

and by Routledge
605 Third Avenue, New York, NY 10158

Routledge is an imprint of the Taylor & Francis Group, an informa business

Funded by Australian Research Council

British Library Cataloguing-in-Publication Data
A catalogue record for this book is available from the British Library

ISBN: 978-1-032-28341-8 (hbk)
ISBN: 978-1-032-28344-9 (pbk)
ISBN: 978-1-003-29637-9 (ebk)

DOI: 10.4324/9781003296379

Typeset in Times New Roman
by MPS Limited, Dehradun

Contents

Acknowledgements

The research on which this book was based was supported by multiple funded projects, including Swedish Research Council (VR) grant no. 2018-05925 and Swedish Research Council for Sustainable Development (Formas) Grant no. 2019–00451. The authors are equally grateful for the support of the Australian government through Australian Research Council Discovery Early Career Researcher Award (DE200100712), which also facilitated the book's open access arrangement. The material on Australia is largely from Sonia Garcia Garcia's doctoral thesis: *Policy disconnections in the regulation of sustainable seafood in Australia* (2019, UTS) for which Kate Barclay was primary supervisor. For personal reasons Dr. Garcia was unable to contribute to this book but gave permission to the authors to use her work. Australian material for the book also came from projects funded by the Fisheries Research and Development Corporation (FRDC project 2014-301, FRDC 2017-092).

Acronyms

AFMA	The Australian Fisheries Management Authority
CCAMLR	Commission for the Conservation of Antarctic Marine Living Resources
CCCIF	Command Centre for Combating Illegal Fishing of Thailand
CCSBT	The Convention for the Conservation of Southern Bluefin Tuna
CFP	Common Fisheries Policy
CDS	Catch Documentation Scheme
CoOL	Country of Origin Labelling (US)
DAFF	Department of Agriculture, Fisheries and Forestry
DG EMPL	Directorate-General for Employment, Social Affairs and Inclusion of European Union
DG MARE	Directorate-General of Maritime Affairs and Fisheries of European Union
DLPW	Department of Labour Protection and Welfare
DoE	Department of Employment of Thailand
DoF	Department of Fisheries of Thailand
EBFM	Ecosystem Based Fisheries Management
EEZ	Exclusive Economic Zone
ESD	Ecologically sustainable development
EU	European Union
FSANZ	Food Standards Australia New Zealand
GSP	Generalised Scheme of Preferences
GVP	Gross value of production
GT	Gross tonnage
ICCAT	International Commission for the Conservation of Atlantic Tunas
ILO	International Labour Organization
IPOA-IUU	International Plan of Action to Prevent, Deter and Eliminate Illegal, Unreported and Unregulated Fishing
IUU	Illegal, unregulated and unreported

MCS	Monitoring, control and surveillance
MTU	Mobile Transceiver Unit
PIPO	Port-In/Port-Out
PSMA	Port States Measures Agreement
RFMO	Regional Fisheries Management Organization
SFPAs	Sustainable Fishing Partnership Agreements
SIDS	Small island developing states
SIMP	US Seafood Import Monitoring Program
Thai-MECC	Thai Maritime Enforcement Command Centre
TIP	Traffick in persons
UNCLOS	United Nations Convention on the Law of the Sea
US	United States
VMS	Vessel monitoring system
WTO	World Trade Organisation

1 IUU Fishing and the Policy Diffusion of the EU-IUU Regulation

Andrew M. Song

Introduction

In the past two decades, a phenomenon known as illegal, unreported and unregulated (IUU) fishing has come to the fore in policy and academic discussions. While the notion that fishing can be done against established rules – circumventing certain reporting requirements and conducted by a party who is not part of a harvesting arrangement – is neither new nor remarkable in the history of fishing, nowadays IUU has taken on added scientific salience, political urgency and economic significance. It is widely accepted that IUU fishing is harmful, and should be stopped, but there are different approaches to tackling it. Acknowledging this, this book argues that major efforts to fight IUU fishing require careful analyses if the goal is to achieve optimal results and avoid unintended consequences. The book draws on the recent experiences of Thailand and Australia to offer an empirical examination of one increasingly prominent solution, i.e. trade-restrictive measures.

IUU fishing deemed a major problem in fisheries, one with significant social, environmental and economic consequences. The extent of IUU fishing is substantial. Annual IUU landings are estimated to be 26 million tons globally, equivalent to one-in-five wild-caught fish, with illicit profits ranging from $10 to $23 billion per year (Agnew et al., 2009; Pew Trusts, 2018; Sumaila, Alder, & Keith, 2006). IUU fishing has a considerable impact on individual fisheries or fishing areas – e.g. IUU fishing caused a 39% reduction in total catches for Commission for the Conservation of Antarctic Marine Living Resources (CCAMLR) in 2000–2001 and over US$330 million annual losses in the Pacific tuna fisheries from 2016 to 2020 (Willock, 2002; MRAG Asia Pacific, 2021). Overall, OECD estimates that IUU fishing effectively amounts to 'the second largest fish producer in the world by value, after China' (Garcia, Barclay, & Nicholls, 2021). IUU fishing is, thus, a serious and direct threat to the health of global fish stocks as well as to the marine ecosystem through the disturbance of habitats and the bycatch of non-target species such as marine

DOI: 10.4324/9781003296379-1

mammals and seabirds. Further, IUU fishing distorts economic competition, creating unfair benefit to perpetrators while disadvantaging those
who fish in accordance with the law. There are also critical social costs
associated with IUU fishing as it can jeopardize the livelihoods and the
food security of legitimate operators and the wider coastal community.
There are other social ramifications as well: IUU fishing is seen as a
linchpin of inadequate labour and safety conditions with increasing
linkages to transnational crime (Belhabib & Le Billon, 2020; Selig et al.,
2022). As the antithesis to sustainable ways of fishing, the concept of IUU
fishing gained the attention of high-level policy makers, as evidenced by
the 2015 formation of the US federal government Task Force on
Combating IUU Fishing and Seafood Fraud (co-chaired by the US
Departments of Commerce and State and made up of a broad range of
other federal agencies), for example. Today, the idea of IUU fishing
amounts to a powerful and all-encompassing narrative that calls for rapid
and robust mitigating action to ensure ocean health and achieve a fair and
thriving fishing future.

   The idea of IUU fishing is frequently linked to a definition offered by
the 2001 FAO International Plan of Action to Prevent, Deter and
Eliminate Illegal, Unreported and Unregulated Fishing (FAO, 2001).
This commonly cited definition was formulated on the experiences of the
Commission for the Conservation of Antarctic Marine Living Resources
(CCAMLR), whose high-seas focus (i.e. a threat to the lucrative
Patagonian toothfish fishery; see Grilly et al., 2015) specified illegal
fishing as an activity conducted by vessels of countries that are parties to
a regional fisheries management organization (RFMO) but operate in
violation of its rules, or operate in a country's waters without permission.
Unreported fishing is defined as a practice whereby caught fish are not
reported or are misreported to the relevant national authorities or the
RFMO. Unregulated fishing is conducted by vessels without nationality
or registration, or flying the flag of states which do not belong to pertinent
fisheries organizations. Therefore, these vessels are rendered unbound by
the rules. In effect, fishing that is referred to as IUU is now generally
understood to include an entire gamut of fishing activities that are in any
way inconsistent with or violating the management measures in force
(Agnew & Barnes, 2004; Le Gallic, 2008).

   This consolidation of the IUU fishing as a worldwide problem affecting
all types of fishing vessels – regardless of their size or gear and both
domestic waters and the high seas – however, has garnered important
criticisms based on its sweeping characterization. What is defined as IUU
fishing in one country or context is not necessarily viewed that way in
other countries or contexts (Leroy, Galletti & Chaboud, 2016). There is a
great deal of variation in how fisheries are regulated in different countries,
not to mention dealing with the many forms of fishing within each

country. Scholars argue that the mainstream construction of 'illegal', 'unreported' and 'unregulated' fishing, and also the categorical use of 'IUU' in an all-inclusive sense, can, for instance, disregard the diversity, legitimacy and sustainability of small-scale fishing practices and their governing systems, generating unfair treatment of them by the agencies working to eradicate IUU fishing (Song et al., 2020). A more sensitive and context-specific understanding of 'I', 'U' and 'U' might be required to reset the direction of the IUU fishing discourse.

The difficulty of estimating the accurate extent and severity of IUU fishing presents a further challenge to the taken-for-granted depiction of the universality of IUU fishing. Despite increased attention and the use of more sophisticated surveillance and analysis tools (e.g. Dunn et al., 2018; Chuaysi & Kiattisin, 2020; Selig et al., 2022), the magnitude of IUU fishing activities at various scales is still difficult to establish (Song et al., 2019). By nature, it is largely a hidden activity that evades formal observations and official statistics. The association with unlawfulness also makes it a taboo (or even dangerous) subject to approach and expose. Questions about how to define and ascertain the extent of IUU fishing, however, do not diminish the need to appropriately confront the present and future threats IUU fishing poses. Doing so will require many different kinds of enquiry, including ones that are more context-driven and case-based, to arrive at a situated, delicate and responsive suite of mitigating actions.

Trade-based approach to tackling the IUU fishing problem

The ways in which the problem of IUU fishing has been tackled are varied, representing different mechanisms in place. There are at least two main approaches to deterring IUU fishing at the national and international levels. The first relies on so-called 'traditional' measures, based on the framework of the United Nations Convention on the Law of the Sea (UNCLOS) (Fujii, Okochi, & Kawamura, 2021; Ma, 2020). Here, the focus is on applying the jurisdictional power of a sovereign state to control the activities of fishing vessels that fly its flag, navigate its waters or enter its ports in accordance with the maritime zones defined in the UNCLOS (e.g. EEZs, territorial sea, regional fisheries management organizations [RFMOs]). Monitoring, control and surveillance (MCS) measures form the core action with which to achieve compliance against fishing vessels (Le Gallic, 2008). By highlighting the obligations of flag and coastal states, this has been a consistent instrument used internationally and domestically from an early formulation (notably, from the 2001 International Plan of Action against IUU fishing [IPOA-IUU]) to more recent initiatives such as the US Interagency Working Group on IUU Fishing involving the US Coast Guard and the US Navy).

The limited capacity of states to exert control over vessels and jurisdictional waters soon became evident, however, leading to only partial success in preventing IUU fishing. Traditional measures are often associated with administrative and political challenges such as the cost of implementation, resource constraints and the problem of multi-country coordination (Le Gallic, 2008; Garcia et al., 2021). In the Gulf of Guinea region, for example, where IUU fishing is rampant (Okafor-Yarwood, 2019), coastal state capacity in the form of naval assets demonstrably had no bearing on countries' ability to curtail IUU activities, including pirate attacks (Denton & Harris, 2021). The difficulties of reinforcing the duties of coastal and flag states through the MCS approach has elicited growing interest in exploring other types of measures aimed at combating IUU fishing, most notably those related to international trade (Leroy et al., 2016).

Trade-based measures are possible because the products of IUU fishing are often traded and transported across multiple jurisdictions (Young, 2016). Fish and fish products are, in fact, the most highly traded food commodity (Asche & Smith, 2010). Thus, controlling import of fishery products based on information about infringing vessels (e.g. IUU vessel lists) or the traded fish itself (e.g. through import documentation, certification or traceability requirements) becomes a crucial action for an import state. This approach has been broadly reflected in the IPOA-IUU,[1] which calls on the need for 'trade-related measures to reduce or eliminate trade in fish and fish products derived from IUU fishing' (Young, 2016; Ma, 2020).

The two most commonly adopted and implemented trade-related measures against IUU fishing are: (1) the prohibition of market access for particular products based on the evidence of IUU fishing operations gathered from catch certification schemes (i.e. shipment-by-shipment embargo) and (2) species-wide (or country-wide) trade sanctions against countries suspected of IUU fishing (i.e. 'non-cooperating states'), where products not resulting from IUU fishing operations are also banned (Ma, 2020; Hosch, 2016).

In addition to the EU and the US, whose unilaterally devised measures have generated much interest in the topic, regional fisheries management organizations (RFMOs) – including the International Commission for the Conservation of Atlantic Tunas (ICCAT), the Commission for the Conservation of Southern Bluefin Tuna (CCSBT) and the Commission for the Conservation of Antarctic Marine Living Resources (CCAMLR) – and also global multilateral environmental treaties such as the Convention on International Trade in Endangered Species of Wild Fauna and Flora (CITES) have all implemented some form of catch certification schemes to combat IUU fishing over the years (Ma, 2020). The main innovation of catch certification schemes is known as the catch documentation scheme

(CDS), which identifies and certifies legally landed catch, and then tracks them through international trade to the end market. The catch certificate is issued to the first buyer of the unloaded catch from a fishing vessel, while a trade certificate is issued by the processing state when the product is re-exported. These are then linked sequentially via their document numbers to ensure traceability along the supply chain (Hosch, 2016).

CDS is relatively simple to police and enforce, but requires the cooperation of multiple authorities – flag states generating catch certificates, port states checking the legality of landings via catch certificates, processing states issuing and validating trade certificates, and end-market states verifying the existence of all valid certificates at importation. With all involved parties performing their duties along the supply chain, CDS would present a stand-alone and real-time mechanism to seal off markets to illegally sourced products (Hosch, 2016). Its effectiveness has been demonstrated in the case of RFMOs eliminating under-reporting and inducing the recovery of several transboundary tuna stocks through multilateral CDS (Hosch, 2016). Generally, CDS can be an effective tool where the most egregious form of IUU fishing is under-reporting and quota overfishing. However, in fisheries where pirate fishing is more rampant, where there are ports and import markets that absorb the IUU catches or where small-scale fishing operations with limited capacity to comply with CDS requirements are greatly involved, CDS is likely insufficient on its own (Hosch, 2016).

Alternately, the second form of trade-related measure, trade sanctions, are more punitive and sweeping in nature as they are applied to countries perceived to be failing in their duty to combat IUU fishing by the party applying the measure. Again, RFMOs provide early examples. The International Convention for the Conservation of Atlantic Tuna (ICCAT) adopted trade restrictive measures on imports of bluefin tuna in 1994 and 1995 against Belize, Honduras and Panama. ICCAT's member countries then applied the recommendations by implementing domestic regulations. For instance, in 1998 the EU introduced an import prohibition of Atlantic bluefin tuna originating from Belize, Honduras and Panama, followed, in 2001, by the import of bigeye tuna originating from Belize, Cambodia, Equatorial Guinea, Saint Vincent and the Grenadines and Honduras. Likewise, in 2001 Japan prohibited the import of Atlantic bluefin tuna from Belize and Equatorial Guinea (Le Gallic, 2008). The import bans were lifted upon the evidence of the country's increasing cooperation with ICCAT with Panama resuming trade in 1999 and Saint Vincent and the Grenadines in 2001 (Le Gallic, 2008; Hosch, 2016). The impact of these trade sanctions (especially when compared to the shipment-by-shipment restriction based on CDS) was said to be drastic, with all tuna exports from the targeted countries plummeting immediately (Hosch, 2016). Having many lucrative markets (i.e. member states) near-simultaneously closed off to the targeted products makes such multilateral trade sanctions an effective action for depressing IUU activities (Hosch, 2016).

Taken together, the trade-related measures activate the economic control function of import states by restricting the import of fish products when the shipments are suspected of being unlawfully sourced or when the third country's vessels are suspected of being involved in IUU fishing. By design, they rely on reducing revenues from IUU fishing (by making it more difficult to monetize catches) and/or increasing both the operating and capital costs of IUU activities (e.g. higher fuel costs, restrictions on goods and services). These strategies are intended to make the operation of IUU fleets economically unviable (Le Gallic, 2008; Leroy et al., 2016).

While the MCS and trade-based approaches are both needed and complementary to each other, analysts have agreed that the trade-based approaches pose distinct economic advantages for states in terms of the cost of implementation. These approaches represent a major policy innovation believed to be capable of effecting large-scale change in IUU fishing behaviour across geographies. Against this backdrop, the uni-lateral policy actions of the EU and US are perhaps most remarkable and have attracted significant scholarly and civil-society attention in recent years. Because these are the two largest seafood-importing jur-isdictions, any trade restrictions that either introduces would weigh heavily on the economics of seafood-producing countries. Overall, the trade-related approach is still fairly nascent and without extensive acceptance by the international community (Ma, 2020). Nevertheless, Hosch (2016, p. 54) states that the implementation of the EU's policy known as 'EU-IUU Regulation' in 2010 and the launch of a US initiative called 'the Seafood Import Monitoring Program' in 2018 'point to the beginning of a proliferation of unilateral documentation schemes'. There are also calls to expand these measures to other major importing countries, such as Japan and China, in order to amplify their impact across the globe (Leroy et al., 2016; Sumaila, 2019; Freitas, 2022).

The EU context and the IUU regulation

This book's main focus is the European Union's Council Regulation No. 1005/2008 of 29 September 2008 (hereafter, the EU-IUU regulation), because it offers a way to understand the potential of such a trade-related policy to scale out to and affect other countries. How non-EU countries themselves perceive, implement or comply with such a trade-restrictive measure aimed at tackling IUU fishing would prove to be an important marker of the effectiveness of this EU policy. This book, therefore, is a timely and unique appraisal of what the EU-IUU regulation may be capable of achieving in the fisheries governance of non-EU countries.

The EU is the world's largest market and importer of fisheries products. In 2007, illegal fisheries imports into the EU have been estimated conservatively at €1.1 billion a year (Commission of the European Communities, 2007). Fittingly, the EU has been the first jurisdiction to implement unilateral trade-related measures to combat IUU fishing. This action was formalized through a law (EC 1005/2008) passed in 2008 and an implementing regulation (EC 1010/2009) adopted in 2009, together comprising a set of rules known as the EU-IUU regulation. This regulation was aimed at strengthening the market state responsibilities for the EU Member States by establishing a common system to control the inflow of IUU fishing. An official EU communication stated its rationale as follows: 'the best way to put an end to this lucrative business is to remove the incentive for crime by making it extremely difficult, if not impossible, to market IUU products at a profit' (Commission of the European Communities, 2007, p. 2). The regulation is also described as creating a 'level playing field' in the global fisheries trade (European Commission, 2020, p. 3). Although this regulation builds on earlier multilateral traceability schemes of RFMOs, used to successfully deal with species such as toothfish and tuna (Agnew, 2000; Garcia et al., 2021), and adopts the internationally agreed upon definition of IUU fishing offered in the FAO's, 2001 International Plan of Action, it also contains features that are distinctive and innovative, albeit experimental.

The scope of the EU-IUU regulation is extensive in that it concerns all marine wild caught harvests (with some exceptions, such as molluscs) landed or imported into the EU, which originate from non-EU flag vessels. As noted above, it consists of a catch certification scheme applying to all traded fish products and a separate but related trade sanction component involving a ban of fisheries imports from countries identified as having unsatisfactory control of IUU fishing by their flag vessels (Hosch, 2016).

When it comes to catch certificates, a competent authority of the flag state is responsible for certifying that catches are being made in accordance with the applicable laws, regulations and international conservation and management measures. The competent authorities would also need to be formally approved by the EU Commission for countries to be allowed to export to the EU (Young, 2016; Ma, 2020). The catch certificate would contain information such as vessel name, fishing licence number, flag state, description and date of catch, and estimated weight of the landings. When a processing state re-exports an imported catch to the EU, a processing statement (or a trade certificate) must be issued at the time of exportation, linking the source products and foreign catch certificates with the end products (Hosch, 2016).

In terms of country-wide import bans, the EU-IUU regulation involves identifying 'non-cooperating third countries' and applying a blanket ban on seafood imports from those countries. The identification process is initiated by reviewing a range of information such as possession of IUU-listed vessels, compliance with RFMO conservation and management measures and adherence to 'assistance requests' made by the EU, in addition to the catch certificate data (Hosch, 2016). If, after notices and consultations, relevant flag states' actions do not result in improved compliance and transparency, the EU may impose trade restrictions including both an import ban into the EU of all fishery products caught by the fishing vessels flying the flag of such states and the non-acceptance of catch certificates accompanying such products (Ma, 2020).

According to this system of rules, the EU began identifying the first batch of countries at the end of 2012. The EU Commission starts the process by initiating consultations with non-EU countries pursuant to Article 51 of the EU-IUU Regulation. Known as 'mutual assistance requests,' country visits may occur by a European delegation that includes the staff of the European Fisheries Control Agency (Hosch, 2016). Based on the findings, the EU Commission may request that the country implement certain changes to improve its control of IUU fishing, and upon successful follow-up actions by the third country, a formal identification (known as the 'yellow card') could be avoided altogether. No bilateral exchanges or reports are publicly available for this part, making it difficult to assess the precise interactions or their effectiveness (Hosch, 2016).

If the EU Commission's mutual assistance requests, resulting from the bilateral dialogue, fail to produce reforms that the EU Commission deems satisfactory, then it will formally pre-identify the country by issuing a 'yellow card'. To justify this decision, the Commission publishes a list of the third country's shortcomings, which serves to demonstrate the country's failure to address IUU fishing (Hosch, 2016). The yellow-carded country is then required to formulate an official plan for the EU Commission, showing how it intends to rectify the situation (Kadfak & Linke, 2021). A formal dialogue process ensues, involving visits by EU delegations to the identified country; this ultimately leads the Commission to either lift the identification (by issuing a 'green card') or apply a formal identification (by issuing a 'red card'), depending on whether the EU is sufficiently assured that IUU fishing is being curtailed. The formal identification as a non-cooperating third country then institutes a trade ban on all fisheries products originating from vessels flagged to that country. Conversely, being de-listed to the 'green card' implies that the third country has suitably addressed and eliminated all identified IUU fishing issues (Hosch, 2016).

As of 20 January 2023, the EU Commission has issued yellow cards to 27 countries, 12 of which were later de-listed without receiving red cards

Table 1.1 Third countries affected by the yellow, red and green cards (status as of 20 January 2023)

	Yellow	Red	Delisted/green	Yellow
Belize	Nov-12	Nov-13	Dec-14	
Curacao	Nov-13		Feb-17	
Panama	Nov-12		Oct-14	Dec-19
Ecuador	Oct-19			
Saint Kitts and Nevis	Dec-14			
Trinidad and Tobago	Apr-16			
Saint Vincent and the Grenadines	Dec-14	May-17		
Guinea	Nov-12	Nov-13	Oct-16	
Togo	Nov-12		Oct-14	
Liberia	May-17			
Sierra Leone	Apr-16			
Ghana	Nov-13		Oct-15	Jun-21
Cameroon	Feb-21	Jan-23		
Comoros	Oct-15	May-17		
Sri Lanka	Nov-12	Oct-14	Jun-16	
Thailand	Apr-15		Jan-19	
Cambodia	Nov-12	Nov-13		
Vietnam	Oct-17			
Philippines	Jun-14		Apr-15	
Taiwan	Oct-15		Jun-19	
South Korea	Nov-13		Apr-15	
Papua New Guinea	Jun-14		Oct-15	
Solomon Islands	Dec-14		Feb-17	
Vanuatu	Nov-12		Oct-14	
Kiribati	Apr-16		Dec-20	
Tuvalu	Dec-14		Jul-18	
Fiji	Nov-12		Oct-14	

IUU-watch, http://www.iuuwatch.eu/map-of-eu-carding-decisions/

(see Table 1.1). Three countries – Belize, Guinea and Sri Lanka – were de-listed only after receiving red cards. Saint Vincent and the Grenadines, Comoros, Cambodia and Cameroon have red cards in place, while six countries currently retain the yellow card status. Panama and Ghana are in an unenviable position of receiving yellow cards for the second time despite being de-listed several years prior, indicating the EU's continual effort in scanning and reviewing the performance of third countries.

The EU stands out as the only jurisdiction unilaterally adopting the country-wide import ban against IUU fishing. Its catch certification scheme is equally sweeping in that the requirement applies to all species, products and volumes (as opposed to targeting certain at-risk species as is the case of the US Seafood Import Monitoring Program SIMP).[2] The EU policy is unique in this way but is not without shortcomings, such as the lack of transparency in the processes leading to identification (see Hosch, 2016), the asymmetrical power relations during negotiations, wherein the EU holds power over the third country (Kadfak & Linke, 2021) and the risk of unfairly targeting small-scale fisheries (see Song et al., 2020). The apparent concentration of EU identifications to Africa, Asia, the Caribbean, and the South West Pacific to date, 41% of which are small island developing states (SIDS), is noteworthy in this regard (see Table 1.1). In comparison, US involvement is more evenly distributed between world regions and also targets advanced fishing nations, including China, Russia, Chile and Canada (NOAA, 2021). Although determining the effectiveness of the still relatively recent EU-IUU regulation is not an easy task, as evidence of actual and measurable reductions in IUU fishing is rare, there are indications that EU identifications have motivated the targeted countries to improve fisheries governance in the form of updated legal frameworks and planning documents (Le Gallic, 2008; Hosch, 2016). Additionally, it has been acknowledged that the EU's identification procedure presents a normative signalling to countries, especially to flag states trading seafood into the EU, that continued failure to address IUU fishing will be subject to heavy scrutiny and will likely result in sanctions. Hence, the EU is spreading the counter-IUU fishing norms, and in doing so, the EU-IUU regulation is enacting two duties: directly advising the target countries on the guidelines for fisheries governance reform and signalling to the world its normative stand of the EU fisheries policy.

Several analysts have suggested that applying economic pressure through EU-inspired trade sanctions could gradually improve the global situation of IUU fishing (e.g. Hosch, 2016; EJF, 2021). There are even direct calls to expand or replicate the trade-based approach the EU has spearheaded to other major market states such as Japan and China to be more effective at sealing off IUU fishing worldwide (Sumaila, 2019, Freitas, 2022). Japan is in an advanced stage of effecting its own rule with the 'Proper Domestic Distribution and Importation of Specified Aquatic Animals and Plants Act', scheduled to enter into force in December 2022. This new law will require records on catches and transfers to be gathered and submitted to the Japanese government in order to establish traceability, including a 'certificate of legal catch' from a foreign government for any imports (MAFF, 2022). With a possibility that worldwide market states normatively evolve in the direction of the EU standards (Leroy et al., 2016), a proliferation of unilateral trade-based measures to combat IUU fishing therefore appears real.

Unanswered questions about the global spread of the EU-IUU regulation

Given the general optimism and growing momentum of the trade-based approach, it becomes necessary to ask to what extent the trade-based measures will succeed in spreading to other countries. How do other countries go about adopting such an approach imbued with (potentially) different normative, administrative and technical bases? Would other seafood export/import nations consider joining the same ambition? Also, for those countries who may face trade sanctions themselves, how receptive or responsive are they to conforming to the new rules? Will their reform efforts be 'deep-rooted' enough, for example, so that compliance efforts are sustained and re-identification avoided (i.e. second issuance of yellow cards to Panama and Ghana)?

These questions matter because the EU-IUU regulation has so far contributed significantly to the trend of pressuring both flag and market states to make visible efforts towards eliminating IUU fishing. While 'upping the ante' through an extended network of similar (trade-related) measures is deemed necessary for a more effective fight against IUU fishing globally, at the same time, they likely mean greater consequences for more countries worldwide in terms of the burden of compliance and threat of trade sanctions. It was reported that SIDS are most vulnerable to seafood trade sanctions, since an average of 7% of all of their exports are fish. SIDS also generate an average of 1.8% of GDP from seafood trade, compared to an average of less than 0.4% for all other types of nations (UNCTAD, 2016). Meanwhile, a recent study showed that tuna fisheries in the Western and Central Pacific Ocean by Pacific Island countries made a substantial reduction in the total annual volume of IUU fishing product (i.e. harvested or transhipped) from 306,440t in 2010–2015 to 192,186t in 2017–2019 (MRAG, 2021). Therefore, it is questionable whether the intensity with which SIDS have been targeted by the EU-IUU regulation is indeed warranted.

Spreading of trade-related measures across the globe can impact countries in another way too. More market countries requesting catch certificates (especially if they are not multilateral or coordinated and hence are different from each other) will mean multiplied administrative and financial burdens on exporting countries. Hosch (2016, p. 54) has warned that 'the burden of complying with complex and decentralised systems (i.e. systems not operating a centralised electronic platform) is particularly difficult for developing countries, their industry and their administrations and, in some cases, this may be beyond their means. In such cases, the technical dimension of proliferating unilateral instruments may reveal itself as a de facto barrier to trade'. As such, developing countries with poor reporting conditions will likely require additional resources to manage these requirements (Leroy et al., 2016).

Lastly, understanding the spread of the trade-related measures will go some way towards informing the discussion of the effectiveness of the EU-IUU regulation. The success of a policy is not determined solely by what it has achieved from the perspective of the initiator. Rather, in a relational way, success is determined by the degree to which the target community accepts the new intervention and in what form, i.e., how they make it their own (for example, see Miller, Bush & Mol, 2014). In other words, the success of a regulation would be something that becomes evident only in the aftermath of others having conformed to it (Song et al., 2019). In addition to the EU's assessment of how the EU-IUU regulation has performed (as demonstrated through the issuance of yellow, red and green cards), a deeper understanding of the non-EU countries who adopt, or are otherwise inspired by, the normative and technical proceedings of the EU-IUU regulation would also be a meaningful contribution to the discussion of markers of success.

The theoretical scaffolding of policy diffusion

This book examines how non-EU (typically termed 'third') countries themselves perceive, implement and/or comply with trade-restrictive measures, of which the EU-IUU regulation is a leading example. Doing so helps us to gauge the potential of this type of unilateral trade-related policy to scale out to and affect IUU fishing outcomes. To organise and gain theoretical support for the analysis, this book relies on what is broadly known as the idea of policy diffusion. Public policy innovations and their diffusion from one jurisdiction to another have emerged as the focus of increasing scholarly attention since the late 1960s (Eyestone, 1977; Gray, 1973; Mörth, 1998; Savage, 1985; Walker, 1969). More recently, policy diffusion has translated into a sizable body of literature on the diffusion of environmental policy instruments with respect to topics such as climate change, pollution and energy policy (e.g. Tews, Busch & Jörgens, 2003; Holzinger, Knill & Sommerer, 2008; Huber, 2008; Biesenbender & Tosun, 2014). For the purpose of this book, policy diffusion is an umbrella term that covers several similar concepts – most notably, policy transfer, policy mobility and policy convergence (Dolowitz & Marsh, 1996; Evans & Davies, 1999; Wolman, 1992; Klingler-Vidra & Schleifer, 2014). Together, policy diffusion encompasses a continuum of exchanges that includes both voluntary and coercive forms of spread that may rely on normative mechanisms such as policy emulation, harmonisation, lesson-drawing as well as structural and political forces involving institutions and incentives (Stone, 2012; Mukhtarov, 2014; Lawless et al., 2020). The push for the diffusion of a new policy can either come from domestic actors or from outside, such as other jurisdictions or international organisations (Shipan & Volden, 2008). Overall, policy diffusion is concerned with a trend of successive or sequential adoptions of a policy spreading outward from a point of origin, where innovation is considered to have arisen (Berry & Berry, 1999; Stone, 2012).

Policy diffusion is theorised according to four main forces: coercion, competition, learning and emulation (Garrett, Dobbin & Simmons, 2008;

Gilardi & Wasserfallen, 2019). Coercion occurs via the pressure from international organisations or powerful countries to adopt certain policies, whereas competition happens when countries enact policies to attract resources and investment over other competing countries. Learning is a mechanism based on analysing the consequence of policies enacted elsewhere. Finally, emulation occurs when policies are enacted by conforming to shared norms and conventions.

Although policy diffusion represents an influential way of thinking about how policy spreads across discrete polities (i.e. countries), a main criticism of the diffusion idea has been the rationalistic, apolitical and mechanistic/positivistic assumptions with which it is implicitly associated. In other words, there is an implicit assumption of general consensus about a policy and that the policy represents an accepted idea 'whose time has come' for wider uptake (Stone, 2012). The assumption here is that a policy is a given that retains its shape and content, and that it is sufficient for another jurisdiction to merely agree to and comply with the same policy. It coincides with seeing policy diffusion as primarily an approach to problem solving, that is, an external solution to be scaled out, contextualized and implemented elsewhere to engineer greater effectiveness (Johnson & Hagström, 2005; Stone, 2012; Mukhtarov, 2014). As argued by others, what is generally missing is any discussion of countries' internal factors, such as the political dynamics, the administrative legacy and the sociohistorical and cultural configurations, which can be more powerful determinants of what gets adopted (and how) than the external policy itself or the modes of spread (Stone, 2012; Steenbergen, Song & Andrew, 2022).

Perspectives more sympathetic to such a social constructivist view have been brought forward highlighting the non-linear, relational and unpredictable reality of policy diffusion. Often phrased as a policy translation, modification of meaning and multiple interpretations of policy ideas are considered inevitable as a policy traverses different contexts (Mukhtarov, 2014). This so-called 'indigenization' of policy can lead to different outcomes than may have originally been envisaged, a transformation into something entirely new, or even a wholesale failure to adopt a particular policy – thus ensuring difference and diversity in the spread of a policy measure (Stone, 2012).

The implication of this nuanced view to analysis of a policy in a global setting is that the success of policy diffusion is dependent on 'other' countries for implementation, not the initiating countries, whose powers and capacity to impose regulations are therefore inherently limited (Johnson & Hagström, 2005; Stone, 2012). These views have usefully challenged the traditional diffusion paradigm and contributed to shifting the focus of a policy spread from sender to receiver.

Two countries' experiences

The main analyses of this book are aligned with this approach. To facilitate an understanding of the potential spread and adoption of a trade-related policy measure globally, this book presents in-depth, primary data-based accounts of the processes through which the EU-IUU regulation has been received and adopted in two non-EU countries (or lack thereof). The two countries in focus, Thailand and Australia, have very different relations to the EU seafood market; Thailand is a major exporter of seafood to the EU, while Australia does not rely on the EU market but is, in fact, a sizable importer of fish on its own. Notwithstanding these differences, they are similar in terms of policy diffusion, in that they are both positioned on the receiving end of the EU-IUU regulation, directly and indirectly. In the chapters that follow, the case of Thailand will be understood through the coercion lens of the policy diffusion theory, reflecting the yellow card issued (and later retracted), which compelled Thailand to undertake domestic fisheries reform. Australia resembles a case of emulation, which may look to devise and apply its own set of a trade-restrictive import policy to pressure exporting countries to reduce IUU fishing. Along the way, the book highlights the context-driven issues, governing configurations and discursive debates in these two countries which in unique ways affect the makeup, trajectory and ongoing outcomes of the outwardly spread of the EU-IUU regulation. Overall, the book offers a rare glimpse into the domestic dynamics associated with policy diffusion; its aim is to inform the global community about what they can expect regarding future implementation of the trade-based regulation as it continues to affect ever more countries.

Notes

1 See Internationally Agreed Market-Related Measures (Articles 65 to 76).
2 There are, however, calls to expand the certification scheme to all species from the 13 species initially covered. See, for example, Oceana (2022) Fishing for trouble: Loopholes put illegally caught seafood on Americans' plates.

References

Agnew, D. J. (2000). The illegal and unregulated fishery for toothfish in the Southern Ocean, and the CCAMLR catch documentation scheme. *Marine Policy, 24*(5), 361–374.

Agnew, D. J., & Barnes, C. T. (2004). Economic aspects and drivers of IUU fishing: Building a framework. In *Fish Piracy: Combating illegal, unreported and unregulated fishing* (pp. 169–200). Paris: OECD Publishing.

Agnew, D. J., Pearce, J., Pramod, G., Peatman, T., Watson, R., Beddington, J. R., & Pitcher, T. J. (2009). Estimating the worldwide extent of illegal fishing.

PLoS ONE, 4(2), e4570. Retrieved February 21, 2023, from 10.1371/journal.pone.0004570

Asche, F., & Smith, M. D. (2010). Trade and fisheries: Key issues for the World Trade Organization (No. ERSD-2010-03). WTO Staff Working Paper.

Belhabib, D., & Le Billon, P. (2020). Illegal fishing as a trans-national crime. *Frontiers in Marine Science, 7*, 162.

Berry, F. S. and Berry, W. D. (1999). Innovation and diffusion models in policy research. In: P. A. Sabatier (ed.), *Theories of the policy process.* Boulder CO: Westview Press.

Biesenbender, S., & Tosun, J. (2014). Domestic politics and the diffusion of international policy innovations: How does accommodation happen? *Global Environmental Change, 29,* 424–433.

Chuaysi, B., & Kiattisin, S. (2020). Fishing vessels behavior identification for combating IUU fishing: Enable traceability at sea. *Wireless Personal Communications, 115*(4), 2971–2993.

Commission of the European Communities (2007). Communication from the Commission to the European Parliament, the council, the European economic and social committee and the committee of the regions on a new strategy for the Community to prevent, deter and eliminate Illegal, Unreported and Unregulated fishing. Retrieved February 1, 2023, from https://eur-lex.europa.eu/legal-content/EN/TXT/?uri=celex:52007DC0601.

Denton, G. L., & Harris, J. R. (2021). The impact of illegal fishing on maritime piracy: Evidence from West Africa. *Studies in Conflict & Terrorism, 44*(11), 938–957.

Dolowitz, D., & Marsh, D. (1996). Who learns what from whom: A review of the policy transfer literature. *Political studies, 44*(2), 343–357.

Dunn, D. C., Jablonicky, C., Crespo, G. O., McCauley, D. J., Kroodsma, D. A., Boerder, K., Gjerde, K. M. and Halpin, P. N. (2018). Empowering high seas governance with satellite vessel tracking data. *Fish and Fisheries, 19*(4), 729–739.

EJF (2021). A race to the top: Lessons learnt from the EU's law on illegal fishing to secure an EU framework to lead global sustainable corporate governance. Retrieved February 21, 2023, from https://ejfoundation.org/resources/downloads/EU-SCG-Policy-Brief-2021.pdf

European Commission (2020). Report from the Commission to the European Parliament and the council on the application of Council Regulation (EC) No 1005/2008 establishing a community system to prevent, deter and eliminate illegal, unreported and unregulated (IUU) fishing (the IUU Regulation).

Evans, M., & Davies, J. (1999). Understanding policy transfer: A multi-level, multi-disciplinary perspective. *Public Administration, 77*(2), 361–385.

Eyestone, R. (1977). Confusion, diffusion, and innovation, *American Political Science Review, 71*(2), 441–447.

FAO (2001). International Plan of Action to prevent, deter and eliminate illegal, unreported and unregulated fishing. Rome, FAO. 24p. Retrieved February 20, 2023, from https://www.wto.org/english/tratop_e/rulesneg_e/fish_e/2001_ipoa_iuu.pdf.

Freitas, B. (2022). Urgent opportunity for EU, US, and Japan to jointly turn the tide on illegal fishing, Environmental Justice Foundation. Retrieved February

21, 2023, from https://policycommons.net/artifacts/2253816/urgent-opportunity-for-eu-us-and-japan-to-jointly-turn-the-tide-on-illegal-fishing/3012513/

Fujii, I., Okochi, Y., & Kawamura, H. (2021). Promoting cooperation of monitoring, control, and surveillance of IUU fishing in the Asia-Pacific. *Sustainability*, *13*(18), 10231.

Garcia, S. G., Barclay, K., & Nicholls, R. (2021). Can anti-illegal, unreported, and unregulated (IUU) fishing trade measures spread internationally? Case study of Australia. *Ocean & Coastal Management*, *202*, 105494.

Garrett, G., Dobbin, F., & Simmons, B. A. (2008). Conclusion: The global diffusion of markets and democracy. In Geoffrey Garrett, Frank Dobbin, & Beth Simmons (Eds.), *The global diffusion of markets and democracy* (p. 344). Canbridge: Cambridge University Press.

Gilardi, F., & Wasserfallen, F. (2019). The politics of policy diffusion. *European Journal of Political Research*, *58*(4), 1245–1256.

Gray, V. (1973). Innovation in the states: A diffusion study, *American Political Science Review*, *67*(4), 1174–1185.

Grilly, E., Reid, K., Lenel, S., & Jabour, J. (2015). The price of fish: A global trade analysis of Patagonian (Dissostichus eleginoides) and Antarctic toothfish (Dissostichus mawsoni). *Marine Policy*, *60*, 186–196.

Holzinger, K., Knill, C., & Sommerer, T. (2008). Environmental policy convergence: The impact of international harmonization, transnational communication, and regulatory competition. *International Organization*, *62*(4), 553–587.

Hosch, G. (2016). Trade measures to combat IUU fishing: Comparative analysis of unilateral and multilateral approaches. International Institute for Sustainable Development and International Centre for Trade and Sustainable Development, Geneva, Switzerland.

Huber, J. (2008). Pioneer countries and the global diffusion of environmental innovations: Theses from the viewpoint of ecological modernisation theory. *Global Environmental Change*, *18*(3), 360–367.

Johnson, B., & Hagström, B. (2005). The translation perspective as an alternative to the policy diffusion paradigm: The case of the Swedish methadone maintenance treatment. *Journal of Social Policy*, *34*(3), 365–388.

Kadfak, A., & Linke, S. (2021). More than just a carding system: Labour implications of the EU's illegal, unreported and unregulated (IUU) fishing policy in Thailand. *Marine Policy*, *127*, 104445.

Klingler-Vidra, R., & Schleifer, P. (2014). Convergence more or less: Why do practices vary as they diffuse? *International Studies Review*, *16*(2), 264–274.

Lawless, S., Song, A. M., Cohen, P. J., & Morrison, T. H. (2020). Rights, equity and justice: A diagnostic for social meta-norm diffusion in environmental governance. *Earth System Governance*, *6*, 100052.

Le Gallic, B. (2008). The use of trade measures against illicit fishing: Economic and legal considerations. *Ecological Economics*, *64*(4), 858–866.

Leroy, A., Galletti, F., & Chaboud, C. (2016). The EU restrictive trade measures against IUU fishing. *Marine Policy*, *64*, 82–90.

Ma, X. (2020). An economic and legal analysis of trade measures against illegal, unreported and unregulated fishing. *Marine Policy*, 117, 103980.

MAFF (2022). Act on ensuring the proper domestic distribution and importation of specified aquatic animals and plants. Retrieved February 21, 2023, from https://www.jfa.maff.go.jp/220614.html

Miller, A. M., Bush, S. R., & Mol, A. P. (2014). Power Europe: EU and the illegal, unreported and unregulated tuna fisheries regulation in the West and Central Pacific Ocean. *Marine Policy, 45,* 138–145.

Mörth, U. (1998). Policy diffusion in research and technological development: No government is an island. *Cooperation and Conflict, 33*(1), 35–58.

MRAG Asia Pacific (2021). The Quantification of illegal, unreported and unregulated (IUU) fishing in the Pacific Islands region – a 2020 Update. 125 p. Retrieved February 21, 2023, from https://mragasiapacific.com.au/projects/quantification-of-iuu-pacific-islands-region/

Mukhtarov, F. (2014). Rethinking the travel of ideas: Policy translation in the water sector. *Policy & Politics, 42*(1), 71–88.

NOAA (2021). Report on the implementation of the U.S. seafood import monitoring program. Retrieved February 21, 2023, from https://media.fisheries.noaa.gov/2021-05/SIMP%20Implementation%20Report%202021.pdf

Oceana (2022). Fishing for trouble: Loopholes put illegally caught seafood on Americans' plates. Retrieved February 21, 2023, from https://usa.oceana.org/wp-content/uploads/sites/4/IUU-21-0013-SIMP-Report_m1_DIGITAL_SPREADS.pdf

Okafor-Yarwood, I. (2019). Illegal, unreported and unregulated fishing, and the complexities of the sustainable development goals (SDGs) for countries in the Gulf of Guinea. *Marine Policy, 99,* 414–422.

Pew Trusts (2018). *The Port State Measures Agreement: From intention to implementation.* Retrieved February 21, 2023, from https://www.pewtrusts.org/en/research-and-analysis/issue-briefs/2018/04/the-port-state-measures-agreement-from-intention-to-implementation

Savage, R. (1985), Diffusion research traditions and the spread of policy innovations in a federal system. *Publius, 15*(4), 1–28

Selig, E. R., Nakayama, S., Wabnitz, C. C., Österblom, H., Spijkers, J., Miller, et al. (2022). Revealing global risks of labor abuse and illegal, unreported, and unregulated fishing. *Nature Communications, 13*(1), 1–11.

Shipan, C. R., & Volden, C. (2008). The mechanisms of policy diffusion. *American Journal of Political Science, 52*(4), 840–857.

Song, A. M., Scholtens, J., Barclay, K., Bush, S. R., Fabinyi, M., Adhuri, D. S., et al. (2020). Collateral damage? Small-scale fisheries in the global fight against IUU fishing. *Fish and Fisheries, 21*(4), 831–843.

Song, A. M., Hoang, V. T., Cohen, P. J., Aqorau, T., & Morrison, T. H. (2019). 'Blue boats' and 'reef robbers': A new maritime security threat for the Asia Pacific? *Asia Pacific Viewpoint, 60*(3), 310–324.

Steenbergen, D. J., Song, A. M., & Andrew, N. (2022). A theory of scaling for community-based fisheries management. *Ambio, 51*(3), 666–677.

Stone, D. (2012). Transfer and translation of policy. *Policy Studies, 33*(6), 483–499.

Sumaila, U. R., Alder, J., & Keith, H. (2006). Global scope and economics of illegal fishing. *Marine Policy, 30*(6), 696–703.

Sumaila, U. (2019). A carding system as an approach to increasing the economic risk of engaging in IUU fishing? *Frontiers in Marine Sciences, 6,* 34. Retrieved February 21, 2023, from 10.3389/fmars.2019.00034

Tews, K., Busch, P.-O., & Jörgens, H. (2003). The diffusion of new environmental policy instruments. *European Journal of Political Research, 42,* 569–600.

UNCTAD (United Nations Conference on Trade and Development (2016). Sustainable fisheries: International trade, trade policy and regulatory issues. Geneva, Switzerland: UNCTAD/WEB/DITC/TED/2015/5. Retrieved February 21, 2023, from https://unctad.org/system/files/official-document/webditcted2015d5_en.pdf

Walker, J. L. (1969). The diffusion of innovations among the American states. *American Political Science Review, 63*(3), 880–899.

Willock, A. (2002). Unchartered Waters: Implementation Issues and Potential Benefits of Listing Toothfish in Appendix II of CITES. TRAFFIC, Cambridge. Retrieved February 21, 2023, from https://agris.fao.org/agris-search/search.do?recordID=XF2015026576

Wolman, H. (1992). Understanding cross national policy transfers: The case of Britain and the US. *Governance, 5*(1), 27–45.

Young, M. A. (2016). International trade law compatibility of market-related measures to combat illegal, unreported and unregulated (IUU) fishing. *Marine Policy, 69,* 209–219.

2 How is Unilateral Trade-Based Policy Adopted and Integrated from the Perspective of Receiving Countries? Applying EU IUU Regulation in Thailand

Alin Kadfak

Introduction

The global crisis caused by unsustainable fishing practices put IUU fishing on the world agenda (Marschke & Vandergeest, 2016; Wilhelm et al., 2020). The European Union (EU) positioned itself as a frontrunner in combatting IUU fishing through the ratification of the Directorate-General of Maritime Affairs and Fisheries (DG MARE) in 2010 (DG MARE, EC Reg no. 1005/2008), hereafter EU IUU regulation. As the biggest seafood market in the world, the EU has exercised its market power on third countries selling seafood products to the EU, requiring importing states to comply with the EU IUU regulation. Since the 1990s, the EU has recognized the potential of using trade in seafood products as a tool to achieve sustainability goals internationally (Thorpe et al., 2022). While the scope of the EU IUU regulation is global, the regulation has been implemented on a country-by-country basis through bilateral dialogue (Miller, Bush, & Mol, 2014). The EU's view on sustainable fishing practices and external fisheries governance is the foundation of EU IUU regulation. Thus far, studies of the EU IUU regulation have focused on a comparison of the regulation with international trade law (Leroy, Galletti, & Chaboud, 2016; Soyer, Leloudas, & Miller, 2017), with little exploration into how the EU IUU regulation works at a bilateral level (See, for example, Elvestad & Kvalvik, 2015; Miller et al., 2014; Rosello, 2017).

While the Australian case (Chapter 3) examines whether trade measures similar to those of the EU, United States (US) and Japan might spread to other market countries, the Thai case illustrates how the EU rules are applied to a producer country that carries both flag state and port state positions (Garcia, Barclay, & Nicholls, 2021). Thailand is a compelling case for exploring the direct impact of EU IUU regulation on the country's fisheries management. Not only did the yellow card exert

DOI: 10.4324/9781003296379-2

direct market pressure on Thailand, which resulted in successful, rapid and significant reforms to fisheries, the issuance of the yellow card also exposed and brought international attention to the complex problem of human rights violations occurring within the Thai seafood industry (Kadfak & Linke, 2021). The Thai case, thus, allows us to unpack the deliberate kind of policy diffusion that occurred at the receiving end of the anti-IUU policy, in comparison to the diffusion by emulation explored in the Australian case.

In the past two decades, Thailand has been a major seafood exporter with an export net worth of nearly 6 billion USD, making up of 20% of Thailand's overall product exports (USDA, 2018). Prior to the reform, the EU was considered the fourth largest market for Thai seafood product. The economic success of Thai seafood exports came with a cost, however. Since the 1990s, Thai fishing fleets had already fished at an unsustainable rate within the country's Exclusive Economic Zone (EEZ). Thai fishing fleets, therefore, started to fish outside the country's EEZ. Fishing in neighbouring countries had been done both legally, with fishing licenses or co-investment with host country companies, and illegally (Derrick et al., 2017). The illegal fishing practices had been associated with labour abuses on fishing boats, due to the fact that unregistered fishing vessels could conceal working conditions from government authorities (EJF, 2015).[1]

Increasingly, the country has faced challenges and criticisms regarding the conditions for migrant workers in many sectors, including fisheries (Chantavanich, Laodumrongchai, Stringer, 2016). Living in a legally grey area, migrant fish workers have experienced poor working conditions, limited access to welfare services from the government and NGOs and physical/verbal abuse (HRW, 2018). Often these workers have been recruited to work on fishing boats against their will (ILO, 2018). Moreover, these migrant workers have struggled to ask for help or leave due to corruption among law enforcement authorities, debt-bondage and the contextual reality of remote fishing at sea (EJF, 2015; Vandergeest & Marschke, 2021). This problems have been picked up by international media and framed as a 'modern slavery' crisis in fisheries, which has aligned with the current anti-slavery movement in global seafood supply chains (Brown et al., 2019; Couper, Smith, & Ciceri, 2015; Stringer, Burmester, & Michailova, 2022; Wilhelm et al., 2020; Yea, 2022; Yea, Stringer & Palmer, 2022). These international pressures coincided with the EU's decision to issue a yellow-card warning to Thailand in April 2015 and to start an official bilateral dialogue to solve the problem.

In order to return to normal status (i.e. receiving a green card), Thailand needed to work in close collaboration with DG MARE to improve its IUU situation. The EU did not publicly state that labour rights in the fishing industry were included in the measures required.

However, labour/human rights have been an underlying agenda included in the bilateral discussion from the beginning. As stated in an EU official document, '*the EU IUU Regulation does not specifically address working conditions on-board fishing vessels, neither human trafficking. Nonetheless, improvements in the fisheries control and enforcement system on IUU fishing may have a positive impact in the control of labour conditions in the fisheries sector*' (European Commission, 2019). Apart from media and political pressures on the EU to act on labour/human rights violations, Thailand had been removed from the EU Generalised Scheme of Preferences (GSP) in January 2015, which meant that the yellow card was the only trade measure left to pressure the Thai government (Kadfak & Linke, 2021; Mundy, 2018).

In response, since the start of the 2015 reform program, the Thai government has officially proceeded with reforms aimed at "*tackling IUU fishing and labour abuses in the fisheries sector*" and has amended national legislation to, at least in part, align with International Labour Organization (ILO) conventions. Therefore, the EU-Thailand IUU dialogue featured a unique element as, in addition to conventional IUU fisheries management regulations, the labour conditions of workers on fishing boats became unavoidable and central to the reform.

This recent fisheries reform is considered to be the most extensive reform Thailand has ever engaged in. The reform reflects the EU's external fishing goal of ending IUU fishing and other normative values attached to what is considered 'sustainable fishing practice'. We examine the case of Thailand as an instructive example that highlights how domestically driven European normative values are interpreted and being integrated into a broader EU external fisheries policy. We argue that studying the EU IUU dialogue allows us to understand how the EU integrates and translates certain normative values, i.e. sustainable fisheries and labour standards, into the discussion. Through this bilateral policy experience, this chapter examines/emphasizes the way in which the EU IUU regulation has come to reflect the emerging concern of labour standards in seafood trade policy (Orbie, 2011).

EU IUU regulation and the issuing of a yellow card to Thailand

Legal aspects

Being the world largest seafood import market, the EU took on the responsibility and a leading role in addressing the IUU fishing problem globally. EC Reg no. 1005/2008 or EU IUU regulation is considered to be the first regulation with applied trade measure to eliminate, deter and prevent IUU fish practices. This regulation sets a trade bar, whereby fishery products stemming from IUU fishing are prohibited entry into the EU market. All traded fisheries products imported into EU member states are required to demonstrate evidence on non-IUU fishing practices. As stated in the EU IUU regulation, section 13 of EC Reg no. 1005/2008: '*seafood have been harvested in compliance with international conservation and management measures and, where appropriate, other relevant rules applying to the fishing vessel concerned, a certification scheme applying to all trade in fishery products with the Community (EU) shall be put in place*' (European Commission, 2009).

The goal of this regulation is to ensure full traceability of marine products that enter the EU market by means of the catch certificate scheme. All coastal, flag, market and port states are expected to comply with the EU catch certification (European Commission, 2009). This means that the flag state has to certify that catches are legitimate during fishing, transshipping and landing, and that the coastal and port states verify the key information of catch certificate as seafood passes through to the EU. Moreover, the EU will share the information regarding vessels engaging in IUU fishing with third countries to prevent those vessels from landing or processing their catches. Lastly, the EU will not accept catch certificate from non-cooperating third countries, including those who have received EU IUU red cards (see paragraph below, Miller et al., 2014, p. 140). To avoid a trade ban, third countries need to commit to applying national and/or international conservation and management measures throughout the entire supply chain, from fishing to packaging.

Commonly, the EU establishes an informal dialogue with third countries on the seriousness of the situation regarding IUU fishing practices. If a third country does not concretely work on the recommendations that the EU has suggested, the EU then issues a yellow card, which is a warning signal to the country to reform. Further inaction (or non-cooperation) might result in a red card or a complete ban of seafood products from that country into EU member states. Introducing the yellow card allows the EU to institute a formal dialogue with the third country to start to rework its fisheries governance towards compliance with international conservation and management measures and IUU fishing elimination.

The implications and interpretation of these mechanisms are important. Broadly, DG MARE, through its IUU unit, establishes an IUU dialogue together with fisheries-related authorities in the third country. The IUU unit is responsible for assessing the situation of the third countries before starting the bilateral dialogue. The bilateral dialogues continue on until the situations of the third countries are 'stabilised', e.g. the Competent Authority for the certification scheme in the third country has control systems in place that the IUU unit recognizes as sufficient. For evaluation, the EU uses their own internal working reports, UN Agency and NGO reports and news media to evaluate third country improvement regarding the IUU situation. Moreover, beyond deskwork, EU IUU unit officers and delegates inspect and observe on site (Kadfak & Antonova, 2021).

The EU applies its power in external territories by encouraging compliance of a 'good legal framework' in the third country. However, what is considered a good legal framework and fisheries management to eliminate IUU fishing is up to the EU's interpretation. The EU claims that the legal framework it has advised the third country to comply with should be adjusted to the context of each country. It also claims that the IUU regulation creates an equal partnership between the EU and the third country to have an open and equal government-to-government conversation on how to solve the IUU problem. However, the EU has never been in a position of symmetrical power due to its great market power and capacity to apply sanctions (Kadfak & Antonova, 2021). This power asymmetry does not mean that the EU can simply require exporting States to do as the EU wants – exporting states exercise their own agency in these relations. However, in 2020 Thailand was among 15 countries, all of which were the least developed and/or developing countries, that reformed their fisheries management systems to according to the EU's objectives. That is, Thailand and 14 other countries aligned their fisheries management according to EU preferences for legal frameworks on international obligations as flag, port, coastal and market states (European Commission, 2009).

Leading to the yellow card in Thailand

The issuing of the yellow card in April 2015 came as no surprise for stakeholders involved in Thai fisheries management. Since the EU IUU regulation became active in 2010, the EU had been actively engaging with third countries to put anti-IUU fishing at the centre of fisheries governance. Problems of overfishing within country EEZs and distant waters (Clark & Longo, 2022), underreported values for catches (Derrick et al., 2017) and unregistered fishing vessels and gears paved the way for the EU to start raising concerns about Thai fisheries. The IUU fishing practices of fishing vessels carrying Thai flags initiated the first informal discussion between the EU DG MARE working group and the Thai Department of Fisheries (DoF) in 2012. However, the final push that lead to the EU's decision to issue the yellow card in 2015 was the infamous modern slavery crisis that had gained the attention of international media (see timeline of the yellow card in Table 2.1). The issuance of the yellow card turned informal talks to a formal ones, setting the stage for an official bilateral dialogue (see further description of EU-Thailand dialogue in Kadfak & Linke, 2021, pp. 4–5).

Table 2.1 Thailand's engagement with the EU IUU regulation and core events in relations to fisheries and labour reforms

2012 onward	DG MARE expressed concerns to Thailand. EU delegates visited Thailand to check on the IUU situation, but no visible improvement resulted
2014 (June)	Traffick In Person (TIP) Report (tier3 – the lowest tier) by the US government
2014 (Second half)	Media stories on Thailand trafficking and 'trash fish' on Thai fishing boats
2015 (April)	Yellow card – warning to ban all seafood products from entering the EU
2015 (1 June)	Thai government adopts EU IUU policy into The Royal Ordinance on Fisheries B.E. 2558 (2015)
2015 (second half)	Lawsuits on human rights in Thai supply chains and repatriation of trafficked fish workers back to their countries
2016	Human Rights Watch sends letter to pressure the EU
2018	Protest from commercial fishing towards proposal to ratify ILO Convention 188 (work in fishing)
2018 (May)	Labour Dialogue is officially signed through an administrative agreement between the EU and Thailand.
2019 (8 Jan)	EU lifts yellow card
2019 (30 Jan)	Thailand ratifies ILO C188. Thailand is the first country in Asia to ratify C188, among the 20 countries that have ratified to date (December 2022)
2019 onward	Continuation of policy implementation and EU observation in Thailand

Methods and data

This chapter is based on analysis of documents (NGOs reports and Thai government policy documents), observations and semi-structured and structured interviews. Fieldwork was conducted from December 2018 to January 2019, February to March 2020 and November to December 2022. Semi-structured interviews were conducted face-to-face and via phone, with 42 key informants between December 2018 and April 2020. The informants include EU and Thai government officials, directors and staff of international organisations, UN agency officers, researchers, local NGO staff, Thai Fisheries Association advisor and members, boat owners, international funders and private actors (for numbers from each category of informants, see Table 2.2). We also conducted 44 structured interviews with migrant fish workers from Ranong fishing harbour between October 2020 and July 2022. Ranong is a one of the major fishing hubs in Thailand, located in the border area between Thailand and Myanmar. Therefore, all of the fish workers we interviewed were Burmese. In the next section, we elaborate how Thailand has taken the EU IUU regulation into the Thai context, before discussing the impacts of the reforms on two key stakeholders – owners and fish workers.

Table 2.2 Categories of key informants

Key informants: semi-structure interviews	Number of interviews
Thai and EU government officers	11
Former EU politicians	2
International NGOs	5
UN agencies	6
Journalist	1
Thai NGOs	10
Private companies	2
International funder	1
Thai fisheries association and boat owners	4
Total of key informants interviews	42
Migrant fish workers: structured interviews	44
Total	86

Thailand fisheries governance 2.0: influences of EU IUU regulation

The core changes in fisheries management

From 2015, Thailand reformed its fisheries regulation to abide by the EU's demands for higher labour standards and traceability mechanisms. Prior IUU fishing in Thailand had lacked accurate information on where fish were caught and in what volume. The lack of vessel registrations and boat-tracking systems further exacerbated the problem. The EU, therefore, argued for stricter monitoring, control and surveillance (MCS), in order to trace seafood from the moment of catch (European Commission, 2009). The requirement to trace fish was the starting point for the Thai fishery reform. The Thai government, therefore, introduced a system which allowed for the identification and tracking of fishing vessels. Many technologies such as a vessel monitoring system (VMS) and Mobile Transceiver Unit (MTU) were introduced to allow vessels to be monitored. VMS, based on satellite technology, became obligatory for all fishing vessels above 30 gross tonnage (GT). Moreover, the Thai government also ordered a complete ban on the operation of all Thai distant fishing fleets following receipt of the yellow card.

The new fisheries law, the Royal Ordinance on Fisheries B.E. 2558, issued in 2015 provided the legal umbrella for the formation and implementation of the Command Centre for Combating Illegal Fishing (CCCIF), an inter-agency taskforce that addresses IUU fishing practices. CCCIF created the Port-In/Port-Out (PIPO) Centres in the coastal provinces. PIPO centre is a multi-authority unit, consisting of a Marine Department (department responsible for ports), Department of Fisheries (DoF), Department of Labour Protection and Welfare (DLPW) and Department of Employment (DoE).

In the first phase of the implementation, the Thai Navy was put in charge of operations to inspect workers, contracts, registration cards, licenses and catch records (Kadfak & Linke, 2021). CCCIF was initiated to focus on the reform of Thai-flag vessels to fish within Thai EEZ. However, CCCIF does not prioritized activities of foreign-flagged vessels fishing outside Thailand's EEZ, but supplying product to Thai processing plants, or transshipping through Thailand. Initially, PIPO centres carried out paper-based inspections, which were time-consuming. Therefore, the so called Fishing Info System, a digitalized fish traceability system, was introduced to replace the paper-based system through which PIPO centres from different government authorities jointly inspect the registration of the fishing boats, logbook of catch at landing and reassure the safety conditions of fishing boats (Kadfak & Widengård, 2022). The Thai government has placed much of the responsibility on boat owners to register fishing vessels, install VMS, apply for commercial use as well as provide documents for individual fish workers. The Fishing Info System connects on-site inspections at the harbours to central control VMS located at the DoF in Bangkok, allowing DoF officers to trace the vessels in real time.

The Thai government received support from Oceanmind, an international NGO with expertise in satellites and artificial intelligence, to apply a machine-learning algorithm to identify suspicious vessel behaviours, in order to monitor and detect high-risk activities. According to our discussion with an Oceanmind representative, high-risk alerts are based on Thai government regulations. These include, for instance, fishing in a closed area, fishing in a licensed area without a license, fishing outside the EEZ, and fishing unlicensed species. Having 31 PIPO centres covering 89 fishing piers in 22 coastal provinces as well as the instalment of VMS on commercial fishing boats reflects the scale of the reform. Putting in place VMS and onsite inspection as the main governing mechanism of tracing fish has also provided a foundation for the Thai government to follow fish workers during fishing trips (for more information on traceability of migrant fish workers, see Kadfak & Widengård, 2022)

Since the yellow card was lifted, CCCIF, which was considered to be a temporary unit dealing with the yellow card, was decommissioned. Since 2020, the work of CCCIF has been transferred to DoF and to the newly established Thai Maritime Enforcement Command Centre (Thai-MECC). Sea inspections became a joint responsibility of three units – Department of Fisheries, Thai-MECC and the Department of Marine and Coastal Resources. PIPO centres continue to monitor and inspect the fishing vessels before and after the fishing trips, with additional activities aimed at supporting Thai-MECC. Thai-MECC has become a focal point for the prevention of IUU fishing, while also ensuring security at sea and other aspects, such as the act of pirate and armed

robbery, terrorism at sea, illegal immigration, accidents and marine res-
cues, forced or slave labour, human trafficking at sea, smuggling of illegal
goods and environment degradation (see more about maritime security
debates in Song, 2021). It is yet unclear which particular aspects of
security at sea Thai-MECC and PIPO will prioritise. However, recent
evidence of online communication, particularly via Thai-MECC and
PIPO Facebook promotional pages, and from our research assistant's
observations on site in Ranong, reveals more surveillance and control
activities, not only on the movement of fishing vessels and the crossing of
trading fleets, but also regarding the cross-border movement of migrant
fish workers between Myanmar and Thailand.

Evolve to something different: EU pressure on Thai labour reform

Human and labour rights problems are a pressing problem for fishing
industries globally. Recently, international advocacy and philanthropist
organisations have problematised labour in fisheries due to the lack of
transparency in seafood supply chains and also regulatory loopholes that
remain (Kadfak, Wilhelm & Oskarsson, 2023). Such pressures are what
influenced the EU to take on labour issues during the dialogue with
Thailand. This taking up of labour issues during a fisheries reform
dialogue in Thailand, in response to the yellow card penalty, represents a
unique case, to date, for EU IUU policy. The EU had initially been clear
that the EU IUU regulation did not include in discussions of human
trafficking within the fishing industry; although it did acknowledge that
'*Different European Commission services as well as the European External
Action Service are working together to tackle the issue of human trafficking and forced labour and share best practices with the Thai authorities*'
(European Commission, 2019). Nevertheless, Thailand was the first
country in Asia to ratify the ILO Protocol of 2014 to the Forced Labour
P029 in June 2018.

At first, the focus on labour reform was towards the criminalisation of
the act of trafficking and forced labour. The US Department of State's
2018 Trafficking in Persons Report criticized the Thai government for
investigating significantly fewer registered cases of labour trafficking in the
fishing industry in 2017 that it did in 2016 (down to 7 from 43). An early
intervention by the Thai government was to established the Ministerial
Regulation on Prevention of Human Trafficking on Labour Operation
Centre in October 2015, under Ministry of Labour, whose aim is to
eliminate all forms of forced labour and improve welfare and working
conditions of workers in the fisheries sector, both on fishing vessels and in
seafood processing factories, as well as to introduce proportional and
deterrent administrative and criminal sanctions. A further critical change
may transpire via the new Ministerial Regulation on Labour Protection in

Sea Fishing Work, B.E. 2561, which was enacted on 26 June 2018. The new law enables labour inspections and criminal proceedings relating to fishing work to be conducted more swiftly and effectively. Since then, the US government also upgraded Thailand from Tier 3 to Tier 2 in the TIP report.[2] This works to improve the perception of Thailand in international arena and increase trust in the country's economic sector, especially for the fishing industry.

The Labour dialogue was formalized in May 2018 between the Thai and EU governments which agreed to discuss the issue formally and to involve core actors such as the and various Thai departments at the Ministry of Labour (MoL). The labour dialogue was a central means to push the issues of recruitment, working conditions and trafficking/forced labour forward, in relation to the fishing sector. Unavoidably, the Labour Dialogue also brought up the challenges of immigration in relation to labour movement and the legal status of migrant workers in Thailand (Boll, 2017). Several pieces of national legislation have been reviewed and amended to ensure an alignment with international standards, resulting in the Emergency Decree amending the Anti-Trafficking in Persons Act BE 2551 (2008) (amended in 2015, 2017, 2019) and the Labour Protection in Fisheries Act BE 2562 (2019), for instance.

The labour reform also introduced inspection, monitoring and traceability mechanisms for individual migrant fish workers on Thai flag vessels. The first means of tracing is focused on migrants' immigration status, where migrant fish workers are now required to become fully documented workers, with some form of official identification. All migrant fish workers are also required to register for a 'seabook' in order to work on fishing boats (see in details Kadfak & Widengård, 2022, pp. 10–11). The seabook is an important first step to registering biometric data, photos and employment records of migrants in Thailand. Another mechanism formalizes workers by connecting work contracts to electronic payment via bank transfer (ILO, 2020). This attempts to replace lump-sum wage payment with monthly salaries, and to replace cash with bank transfers via an ATM card. This way, the Thai government can trace monetary transfers, ensuring that the agreed-upon wages are paid, thereby avoiding debt bondage, which is one form of forced labour. Only after migrants have been registered and have received all mandatory documents and a health card are they allowed to board fishing boats.

Labour inspection at port, aligned with vessel inspection, has been assigned to PIPO. Harbour inspection is supposed to take place before and after every fishing trip, and inspectors are expected to use a biometric face scan system to verify that each individual fish worker matches their registered photo. The detailed information on immigration status and work contracts are also supposed to be double-checked and signed off by the PIPO local official director. Without full authorization from these four authorities – the Marine Department, Department of Fisheries (DoF), Department of Labour Protection and Welfare (DLPW) and Department of Employment (DoE) – fishing trips cannot embark or return. DLPW inclusion in the governing mechanism is significant as it establishes the connection between individual workers and the particular fish stock caught at sea (Kadfak & Widengård, 2022).

Labour traceability allows the Thai government to follow fish workers beyond the fishing trips to the country of origin. This tracing shows an attempt to legalize the recruitment process, which is considered to be the root cause of trafficking (EJF, 2018). This includes, for instance, a legal recruitment pathway for state-to-state memorandums of understanding (MOU), whereby the Thai government signs a contract with source country governments with assigned recruitment agencies. This tracing tries to bypass the informal brokers in the recruitment cycle, who often create a debt-bond situation for migrant workers entering the workplace. The MOU mechanism so far is still a work-in-progress. This is because recruitment via the MOU process has not been popular. In 2019, 69% of new fish workers were recruited via networks of family and friends (ILO, 2020). Our informants mentioned that employers often send the head of the migrant fish workers group (Burmese nationality, in our case) on each fishing boat to go back to their hometowns in Myanmar to recruit more workers. MOU workers often come with a guaranteed job and a contract. Many boat owners do not end up recruiting MOU fish workers. This is because many of the MOU workers appear to lack sufficient skills to be working on fishing boats, which is a dangerous occupation. For example, boat owners whom we interviewed, mentioned that some of MOU fish workers had never experienced living on fishing boats before, so they ended up leaving the sector. Moreover, MOU workers are considered to be more expensive than workers hired through direct recruitment due to the cost of paperwork and formal recruitment agencies involvement.

Box 2.1 Highlight of the main regulatory amendments and implementations

The Royal Ordinance on Fisheries B.E. 2558 (2015) (major amendment after 68 years). This regulation discusses issues that align with EU IUU regulations on:

- Monitoring, control and surveillance
- Traceability
- Elimination all forms of forced labour and improved welfare and working conditions

- New Centre dealing with IUU issue: the Command Centre for Combating Illegal Fishing (CCCIF) in 2015
  - 32 Port-In/Port-Out (PIPO) Centres
  - Vessel Monitoring Systems (VMS) of vessels more than 30?tonnes
- The Ministerial Regulation on Labour Protection in Sea Fishery Work in 2014, and amended in 2018
- Thailand ratified ILO C188: protecting the living and working conditions of fishers on board vessels in 2019
- Continued discussion between ILO and Thai government on ILO Conventions Nos. 87 and 98, on Freedom of Association and the Right to Organise and Bargain Collectively.

While the EU and the ILO have played a significant part in improving labour standards in Thai regulations within the larger fisheries reform, other non-state actors, such as Thai and international NGOs, as well as donors, also contributed to elevating labour standards through advocacy campaigns and private auditing (EJF, 2013, 2015; HRW, 2018; Issara, & IJM, 2017). In particular, NGOs have been fulfilling two roles since the start of the reform. First, they took on a new watchdog role to ensure state and market actors in the supply chains are held accountable for their actions on human and labour rights. For instance, the Thai CSO Coalition, which newly emerged during the seafood slavery scandal, offers a direct strategy for holding one-on-one dialogues with major Thai seafood processing companies to improve conditions for workers and the representation of different nationalities in factory welfare committees. Additionally, these NGOs have now taken on a new partnership role with the private sector. For example, two local NGOs from Sumut Sakorn, a province known as the country's seafood processing hub, have been working as third parties to receive grievances from migrant workers, and bring these issues to the factory board (Kadfak et al., 2023).

The Thai government's fisheries and labour reforms and NGO interventions have brought drastic changes to Thai fisheries. According to the most recent fieldwork, most stakeholders mentioned that migrant fish workers on fishing boats and migrant workers in seafood processing factories are now the most documented and regulated sectors of migrant workers in the country. This is reflected in the recent information provided by DoF that '100% of the migrant workers employed in the fishing and seafood sectors have entered Thailand through legal channels or were approved under the proof of nationality measures' (Department of Fisheries, 2022). The overfishing problem has improved as well. According to one study, the fish catch in 2017 in Thai waters was mostly lower than the maximum sustainable yield point, except for the pelagic fish in the Andaman Sea (Kulanujaree et al., 2020). However, there are some critiques of the rapid, top-down approach of the reform, which is discussed through a policy-diffusion lens in Section 6 below. In the next section, we first discuss how the reform bought new challenges to key stakeholders, including boat owners and fish workers.

Impacts of the reform

The EU IUU regulation applies to all four types of flag, port, coastal and market states. Thailand's seafood supply chains are complex, involving all four types of state measures. For instance, Thailand is one of the tuna capitals on the world through importing frozen tuna and processing and repackaging it for export to major markets like the EU. While we acknowledge that various supply chain actors have been impacted negatively by the modern slavery scandal and the yellow card, this study does not extend to seafood processing companies, brand companies and retailers companies. In this section, we focus on two main actors, the boat owners and migrant fish workers, who have been impacted directly by the fisheries and labour reform.

Boat owners and fisheries associations

Boat owners are the primary group of actors responsible for complying with the reform. During the reform, however, this group was largely excluded from the dialogue between the Thai government and the EU. We have interviewed several boat owners, members of the Ranong Fisheries Association, and one advisor to the Thailand Fisheries Association, who have discussed at length how the reform happened so rapidly, and how they had very few opportunities to provide input to the reform. Negative impacts can be categorized in three ways.

The primary concern of the boat owners was the cost of adopting the new requirements. As discussed in the previous section, the reform to make seafood catch and labour legal and traceable came with a high cost. Boat owners are required to declare and register all of their fishing vessels. They are also responsible for installing VMS and paying the monthly cost of GIS services. Many of the old fishing vessels did not pass the new standards or failed to register the license. The reform has also put stricter rules on the national fishing fleets carrying the Thai flag. There are no official statistics on the number of fishing boats that were banned because of this, but according to an estimate from our informant, 3,000 international fishing vessels carrying the Thai flag faced a complete ban since the beginning of the reform, which has made a major impact on the economy. Thailand's entire commercial fishing fleet was reduced from 25,002 in 2015 to 10,376 in 2020 (EJF, 2022, p. 29). According to a recent study, around 60,000 people, both Thai and migrant workers, lost their jobs due to the high cost of registration and documentation requirements following the reform, which pushed many boat owners to shut down their operations (Wongrak et al., 2021, p. 10).

The reform also forced boat owners to change the way they pay their crew – from a lump sum paid after the sale of the catch, which takes into account the often many months of working and the agreed upon share of the sell, to a monthly salary (Vandergeest & Marschke, 2021). This major change met with much resistance from boat owners, one of whom voiced his disagreement with the new law this way:

> *Seafood prices have been down 30%, and then we have to pay about 30–40% increased costs. What are the increased costs? What about the labour costs? What are the expenses? Before, we used to pay labours a daily wage. So if I go out fishing for 10 days, then I only pay for the days that workers are on the fishing boat. With the new law, we have to pay monthly. So we have to pay when they rest! Of course, when the boat is under repair, we pay workers anyway, because we have to pay them to be able to keep them.*

Second, the complexity of the revised regulations and implementation has been burdensome and confusing for the boat owners. Throughout the five-year reform period, there have been several sub decrees and announcements/notifications that branch out from the main Royal Ordinance on Fisheries, B.E. 2558 (2015) that boat owners are obliged to follow.[3] One boat owner we talked to at Ranong harbour expressed that:

> *There is a lot of confusion in the multiple and complex regulations. For instance, we were not sure what kind of vessel registrations we should follow. We were asking our peers, who also have very little knowledge about the new regulations. For example, if I have a purse seiner, I should not register the vessel as a trawler, but in fact, we could register it without identifying which type of boat it is.*

Boat owners addressed the problem of mounting documentation and digitization processes by hiring additional administrative staff to handle registration paperwork for both boat and fish workers during the reform. The new regulations introduced several new procedures regarding the hiring of fish workers on fishing boats. To tackle the debt-bondage situation, the new regulation specifically asks boat owners to pay migrant fish workers via bank payment. This is to ensure transparency and accountability of a fair minimum wage. However, in practice, this payment method has been burdensome and costly for both boat owners and fish workers.

Third, the reform and the concurrent and infamous scandal of modern slavery reshaped the image of boat owners as the 'bad guy'. They have been portrayed as mafia, criminals or thieves in Thai and international media platforms and NGOs reports. For instance, NGO reports and investigative documentaries have depicted criminal activities whereby fish workers had been deceived and captured on international fishing fleets (EJF, 2013, 2015). This type of a blame game, however, may not create long-term solutions for the reform, as one of our informants mentioned:

> *The head of the CCCIF (at the time) told us that 'we invited you to listen, not to speak. Vessel owners are robbers!' They look at us as thieves!*
> *(Advisor, Fisheries Association of Thailand)*

Framing boat owners as the 'bad guy' has done little to solve the structural problem of corruption that exists as part of the Thai administrative government (Kadfak & Widengård, 2022). Prior to the reform, legal loopholes, myopic immigration policies and a lack of labour rights had all helped employers control of the freedom of fish workers during fishing trips and at the harbour (Vandergeest & Marschke, 2020).

Fish workers on Thai fishing fleets[4]

We conducted interviews with 35 migrant fish workers (fishing crew) based in Ranong. We asked 17 of them specifically about their perceptions and experiences of the recent reform. A majority of fish workers experienced positive changes from the reform (for more discussion regarding the dialogue process from the perspective of government officials, see Kadfak & Linke, 2021). Many mentioned safety improvements, guaranteed monthly payments and a decrease in harassment and abuse from employers. Furthermore, following the reform, two key factors that helped reduce the potential for abuse and violations on fishing boats were (1) the introduction of a 30-day limit on fishing trips and (2) harbour inspections at the commencement and conclusion of fishing trips. For example, one informant explained that

> *I think it's very nice to have legal protection for the fish workers. Working on a boat is a very risky job. When there was no protection law, then there was no fear [of consequences]. If one was killed and dumped in the middle of the sea, no one would have known except the crew. But since the law started to give protection, killing or abuse probably won't happen again.*

The introduction of PIPO inspections at the start and conclusion of fishing trips reduces some risks for fish workers. According to an interview with PIPO, the most common risk in relation to fish workers is a lack of water and food trips that take longer than expected. During inspections, PIPO often pays attention to all food/drink and medical supplies on board. Moreover, an interview with the chief of the VMS workgroup, fishing and fleets management division at DoF, revealed that the centrally coordinated real-time monitoring system may observe certain patterns of fishing routes that may signal alarm of force labour. DoF can inform PIPO at a particular harbour to call a particular vessel in question back for further inspection.

While harbour inspections have increased, there has been an issue of trust among fish workers towards Thai authorities. Working conditions on fishing boats remain problematic despite the reform. This is because harbour inspections have focused on documentation and head counting, rather than on engaging in conversations and/or investigating sleeping/ working arrangements and safety practices on fishing trips (Kadfak & Widengård, 2022).

These inspections have both positive and negative impacts on fish workers. From their perspective, inspections do help ensure their chances of returning safely from fishing trips. At the same time, however, they are time-consuming and provide fish workers less flexibility in seeking jobs in fisheries. Interview respondents informed us:

> *I think PIPO coming to check at the harbour can be both good and bad. The good thing is that they (PIPO) will come and check fish workers. And those who have no proper documents will not be allowed to go on fishing boats, and the fishing trip is then cancelled. The bad thing is that it takes time to check the documents, which makes things difficult when we are trying to leave for fishing.*
>
> *PIPO creates a difficulty regarding required documentation. We cannot just show our passport and jump into the boat. Fish workers must register at PIPO before the fishing trip.*

After the reform, fish workers needed to show identification documents (i.e. certification of identity, pink card, travel document) and the seabook. Most of our informants confirmed that they do not have access to the real documents, only to copies of them (for more details, see Kadfak & Widengård, 2022). The concern that this lack of ownership and possession of one's legal documents may lead to forced labour is not a new. The ILO had discussed this issue prior to the reform. But our findings confirm a lack of improvement on this issue (ILO, 2020). For instance, we learnt that boat owners continue to hold all the original documents and give only copies to the workers. In practice, this means that fish workers are unable to leave or find new employers without informing their current employer. Little is known regarding the actual implications of document bondage, and further study is required. During the group discussion, boat owners maintained that the cost of these documents is very high, so they do not trust fish workers to carry them around during fishing trips or on land.

Two emerging problems regarding the increasing demand for documentation are the cost and accessibility of documents. Some fish workers complained that the additional costs associated with documentation have been transferred from employers to fish workers, and many of those we interviewed mentioned that these costs have been deducted from their salaries.

Policy diffusion of the EU IUU regulation in Thailand

The EU uses the EU-IUU regulation to push the sustainability agenda in marine governance globally. The EU has recognised Thailand as a champion in integrating anti-IUU policy into its domestic regulations.[5] The Thai government also claimed success after four years of rapid reform by taking a leading role in combating IUU fishing in Southeast Asia (Auethavornpipat, 2017; Kadfak & Linke, 2021). Nevertheless, stakeholders have raised many challenges to the so-called sustainability and fairness of this top-down approach of the reform. In this section, we employ a policy-diffusion lens to explore what fell between the cracks during the EU-led reform in Thailand.

First, the EU applied the trade-restrictive IUU regulation in order to bring exporting countries' management of their domestic fleets in line with EU policy and ideals. In other words, the EU works to create a level playing field for all seafood products entering the EU by forcing all EU member states to comply with the same sustainable governing measures.

Therefore, the EU IUU regulation – effected through the carding system and threat of an import ban – allows the EU to control the sustainability of catches outside its jurisdiction and to make seafood products traceable before arriving at the EU's border. This idea of a level playing, however, has not been applied equally to all exporting countries; it affects only those that the EU has defined as problematic or 'non-cooperative'.

To fully understand the logic behind the EU's carding decisions likely requires further study. Still, we can learn from the Thai case that the dialogue that takes place during the carding period become the space for two governments to 'negotiate and tailor' which aspects of IUU fishing are of greatest concern (Kadfak & Linke, 2021). Without a clear set of standardized procedures, the EU IUU delegation became a technical knowledge broker to interpret IUU policy implementation (Lavenex, 2008; Lavenex & Schimmelfennig, 2009). The EU team has been the key actor to evaluate the improvement of the reform. In the case of Thailand, the EU prioritized certain aspects of IUU fishing over the others. More specifically, the EU prioritized strict monitoring, control and surveillance of the harvesting node (by flag states) and paid less attention to coastal and processing states within the supply chains. This aspect of policy diffusion is important and requires further exploration of the underlying reasons why the EU focused on flag state reform over other types of states involved in the seafood trade. This matters because this focus represents only a portion of the seafood caught via Thai fishing fleets that ends up in the EU market, while the majority of domestic catches are for domestic consumption and the Asian market. Yet, Thai fishing fleets have been the main target of the reform, as we elaborate in section Thailand fisheries governance 2.0: influences of EU IUU regulation of this chapter.

Second, the EU IUU regulation is explicitly about fishing practices that contravene rules put in place to protect fish stocks. However, the Thai case illustrates what a non-harmonized policy diffusion of the EU IUU regulation looks like by introducing labour/human rights as an essential part of the reform. In other words, the labour add-on depicts a certain form of policy translation, whereby the initial policy intervention creates multiple interpretations subject to local concerns, which provides different outcomes in the end. Human and labour rights is an emerging topic of policy study in fisheries. What we observe here is the unpredictable and non-linear nature of policy diffusion wherein the use of trade measures to tackle conventional protections of fish stock is intermingled with the diffusion of human rights into natural resource management. While the goal of protecting fish stock is explicit and has been agreed upon internationally, the protection of human/labour rights within fisheries has not yet explicitly been accepted internationally. The EU, therefore, has partnered with the ILO to translate labour-rights protection on fishing fleets via the ratification of ILO C188. The critique remains, however, that many EU member states, including major fishing states such as Spain, have not yet ratified this convention.

Policy diffusion encompasses how political force initiates and drives certain agendas forward. Media exposure of the modern slavery scandal within the Thai fishing industry helped raise the concern of EU market actors, including Global North retailers, consumers' organizations and member state politicians as well. One of our informants, a former member of the EU parliament, told us that powerful images of trafficked fish workers, circulated in international news media, that connect slavey to seafood to the EU market, and ultimately to consumers, have driven the EU to employ urgent action. The media exposure of modern slavery in Thailand should not be viewed apart from this global context. Observations within fisheries align with the maturing modern-slavery framing in resource extraction governance globally (Brown et al., 2019). Such framing has been promoted through large philanthropical organizations and international NGOs attempting to expand consumerism and ethical awareness into supply-chain governance (Kadfak et al, 2023).

Third, EU-led fisheries and labour reforms in Thailand are considered successful in the eyes of the EU. However, policy diffusion is arguably dependent on 'the receiving' countries for implementation (Stone, 2012). Although the Thai government took on the yellow card as a national agenda, regulatory reform has not been an inclusive process, but rather a somewhat brute response to the national emergency agenda. Therefore, many key actors – e.g. commercial fisheries associations, small-scale fisheries associations, seafood companies and migrant workers representatives – often through local NGOs, were not invited to give input on the new regulations. Having a military junta and martial law during the time of reform helps explain these outcomes. Exclusion of the several key stakeholders in the reform raises concern about the sustainability of policy implementation, as voiced by interviewed boat owners: *What do you mean by sustainability? Who is sustainable? Nature can survive, but humans cannot survive … This is not sustainable!* The rapid reform clearly missed out on the situated sociopolitical conditions of diverse actors and institutional settings in fisheries, which in the end could leave a lasting negative impact on the adoption of the new policy in the country (Steenbergen et al., 2022; Stone, 2012).

The question of policy sustainability is an important one. The lifting of the yellow card may lead one to assume that the IUU policy has been fully adopted in the Thai context. However, it is not easy (if not impossible) to find a completion point of policy diffusion when observing through everyday implementation. The lack of agreement from various stakeholders regarding the legitimacy of the new rules creates everyday resistance. For example, the Fisheries Association has negotiated minor illegal activities with provincial government authorities in order to circumvent new regulations imposed by the central government. Since the lifting of the yellow card, the EU and the Thai government have officially established a 'working group' that meets twice a year to follow up ongoing progress. To date (December 2022), the EU continues to request inspections at fishing harbours twice a year to keep pressure on the Thai implementation post the carding system. During our recent fieldwork in November–December 2022, Thai-MECC and PIPO continue to be present and active in enforcing harbour inspections. However, inspections have mostly devolved into documentation check-ups rather than a genuine investigation of working conditions.

Conclusion

Thailand provides a good example of what happens in a country on the receiving end of coercive and deliberate policy diffusion. A policy-diffusion lens helps us to understand how the carding system, as part of the EU IUU regulation, opens up the policy space for diffusion and spread to occur, and it points to the importance of contextualisation. An important lesson is that the EU IUU regulation, despite its economically forceful nature, should not be understood as a policy package bound for straightforward adoption by the receiving country. Instead, IUU fishing refers to specific problems in a specific country; this calls for a non-singular approach to what implementation will look like in each affected country. The problems of labour rights violations and lawless practices of domestic fishing fleets ended up being included in the anti-IUU agenda show EU's influence of EU in the third country domestic reform. The labour add-on is an important empirical contribution the Thailand case prominently offers. It thus leads to a widely applicable question: Should the narrative of IUU fishing continue without the inclusion of labour rights of fish workers? What is gained and/or lost from adding labour rights into anti-IUU policy globally?

Notes

1 Thai fishing fleets, however, did not contribute the raw material to canned tuna processing factories in Thailand. Thai processing companies, in theory, should have been held responsible for flagged abuses on vessels of other countries that supply Thai factories.
2 Tier 3 of the TIP report refers to minimum standards outlined in the Trafficking Victims Protection Act, while Tier 2 refers to countries that makes significant efforts to comply with the standards.
3 https://leap.unep.org/countries/th/national-legislation/royal-ordinance-fisheries-be-2558-2015
4 Seafood processors and migrant fish workers make up two thirds of the workforce in the Thai fishing and seafood processing industries. Pressure resulting from the discourse of modern slavery influenced a major labour reform within the Thai seafood industry. It is important to note that migrant workers within Thai fisheries are divided into two groups: fish workers and seafood processing workers. These two groups have different demographics, recruitment channels, patterns of mobility, document requirements and social support systems (ILO, 2020; Vandergeest & Marschke, 2021). In this chapter, we only focus on the fish workers who are working on the fishing boats.
5 Observed from EU press-release webpages and the EU's 10-year anniversary of the European Union's pioneering EU IUU Regulation webinar, co-organised by EU and the EU IUU Coalition (https://www.iuuwatch.eu/2021/01/event-summary-fighting-iuu-fishing-the-eus-vision-for-healthy-oceans/).

References

Auethavornpipat, R. (2017). Assessing regional cooperation: ASEAN states, migrant worker rights and norm socialization in Southeast Asia. *Global Change, Peace & Security, 29*(2), 129–143.

Boll, S. (2017). Human trafficking in the context of labour migration in Southeast Asia: The case of Thailand's fishing industry. In Ryszard Piotrowicz, Conny Rijken & Baerbel Uhl (Eds.), *Routledge Handbook of Human Trafficking* (pp. 68–77). Routledge: London.

Brown, D., Boyd, D. S., Brickell, K., Ives, C. D., Natarajan, N., & Parsons, L. (2019). Modern slavery, environmental degradation and climate change: Fisheries, field, forests and factories. *Environment and Planning E: Nature and Space, 4*(2), 191 – 207. DOI:10.1177/2514848619887156

Chantavanich, S., Laodumrongchai, S., & Stringer, C. (2016). Under the shadow: Forced labour among sea fishers in Thailand. *Marine Policy, 68*, 1–7.

Clark, T. P., & Longo, S. B. (2022). Global labor value chains, commodification, and the socioecological structure of severe exploitation. A case study of the Thai seafood sector. *The Journal of Peasant Studies, 49*(3), 652–676.

Couper, A., Smith, H. D., & Ciceri, B. (2015). *Fishers and plunderers: Theft, slavery and violence at sea.* London: Pluto Press.

Derrick, B., Noranarttragoon, P., Zeller, D., Teh, L. C. L., & Pauly, D. (2017). Thailand's missing marine fisheries catch (1950–2014). *Frontiers in Marine Science, 4*, 402.

Department of Fisheries (2022). Thailand's progress on combating IUU fishing and labour issues towards fisheries sustainability. Retrieved February 16, 2023, from https://www4.fisheries.go.th/dof_en/view_news/426

EJF (2013). *Sold to the Sea: Human Trafficking in Thailand's Fishing Industry.* Retrieved November 11, 2022, from https://ejfoundation.org/reports/sold-to-the-sea-human-trafficking-in-thailands-fishing-industry

EJF (2015). *Pirates and slaves: How overfishing in Thailand fuels human trafficking and the plundering of our oceans.* Retrieved November 11, 2022, from https://ejfoundation.org/reports/pirates-and-slaves-how-overfishing-in-thailand-fuels-human-trafficking-and-the-plundering-of-our-oceans

EJF (2018). Thailand's Progress in Combatting IUU, Forced Labour & Human Trafficking: EJF Observations and Recommendations

EJF (2022). *Driving improvements in fisheries governance globally: Impact of the EU IUU carding scheme on Belize, Guinea, Solomon Islands and Thailand.* Retrieved November 5, 2022, from https://www.iuuwatch.eu/2022/03/driving-improvements-in-fisheries-governance-globally-impact-of-the-eu-iuu-carding-scheme-on-belize-guinea-solomon-islands-and-thailand/

Elvestad, C., & Kvalvik, I. (2015). Implementing the EU-IUU regulation: Enhancing flag state performance through trade measures. *Ocean Development & International Law, 46*(3), 241–255.

European Commission (2009). Handbook on the practical application of Council Regulation (EC) no. 1005/2008 of 29 September 2008 establishing a community system to prevent, deter and eliminate illegal, unreported and unregulated fishing (the IUU Regulation, MareA4/PSD(2009)A/12880), 1–87. Retrieved February 16, 2023, from http://nafiqad.gov.vn/Portals/0/DOCUMENTS/handbook-original-en.pdf

European Commission (2019). Questions and Answers – Illegal, Unreported and Unregulated (IUU) fishing in general and in Thailand. Retrieved November 5, 2022, from https://ec.europa.eu/commission/presscorner/detail/en/MEMO_19_201.

Garcia, S. G., Barclay, K., & Nicholls, R. (2021). Can anti-illegal, unreported, and unregulated (IUU) fishing trade measures spread internationally? Case study of Australia. *Ocean & Coastal Management, 202*, 105494.

HRW (2018). *Hidden chains: Rights abuses and forced labor in Thailand's fishing industry*. Retrieved February 16, 2023, from https://www.hrw.org/report/2018/01/23/hidden-chains/rights-abuses-and-forced-labor-thailands-fishing-industry

ILO (2018). *Baseline research findings on fishers and seafood workers in Thailand*. Retrieved November 5, 2022, from https://www.ilo.org/asia/publications/WCMS_619727/lang–en/index.htm.

ILO (2020). *Endline research findings on fishers and seafood workers in Thailand*. Retrieved November 2, 2022, from https://www.ilo.org/asia/publications/WCMS_738042/lang–en/index.htm.

Issara, & IJM. (2017). *Not in the same boat: Prevalence & patterns of labour abuse across Thailand's diverse fishing industry*. Retrieved October 28, 2022, from https://www.issarainstitute.org/_files/ugd/5bf36e_367d99d9de2f4c8393-f91ab55d30b374.pdf.

Kadfak, A., & Antonova, A. (2021). Sustainable networks: Modes of governance in the EU's external fisheries policy relations under the IUU Regulation in Thailand and the SFPA with Senegal. *Marine Policy, 132*, 104656.

Kadfak, A., & Linke, S. (2021). Labour implications of the EU's illegal, unreported and unregulated (IUU) fishing policy in Thailand. *Marine Policy, 127*, 104445.

Kadfak, A., & Widengård, M. (2022). From fish to fishworker traceability in Thai fisheries reform. *Environment and Planning E: Nature and Space*, DOI:10.1177/25148486221104992.

Kadfak, A., Wilhelm, M., & Oskarsson, P. (2023). Thai Labour NGOs during the 'Modern Slavery' Reforms: NGO Transitions in a Post-aid World. *Development and Change*. 10.1111/dech.12761

Kulanujaree, N., Salin, K. R., Noranarttragoon, P., & Yakupitiyage, A. (2020). The transition from unregulated to regulated fishing in Thailand. *Sustainability, 12*(14), 5841.

Lavenex, S. (2008). A governance perspective on the European neighbourhood policy: Integration beyond conditionality? *Journal of European Public Policy, 15*(6), 938–955.

Lavenex, S., & Schimmelfennig, F. (2009). EU rules beyond EU borders: theorizing external governance in European politics. *Journal of European Public Policy, 16*(6), 791–812.

Leroy, A., Galletti, F., & Chaboud, C. (2016). The EU restrictive trade measures against IUU fishing. *Marine Policy, 64*, 82–90.

Marschke, M., & Vandergeest, P. (2016). Slavery scandals: Unpacking labour challenges and policy responses within the off-shore fisheries sector. *Marine Policy, 68*, 39–46.

Miller, A. M. M., Bush, S. R., & Mol, A. P. J. (2014). Power Europe: EU and the illegal, unreported and unregulated tuna fisheries regulation in the West and Central Pacific Ocean. *Marine Policy, 45*, 138–145.

Mundy, V. (2018). The impact of the EU IUU Regulation on seafood trade flows: Identification of intra-EU shifts in import trends related to the catch certification scheme and third country carding process. *Environmental Justice Foundation, Oceania, The Pew Charitable Trusts, WWF. Brussels, Belgium.* Retrieved February 16, 2023, from https://europe.oceana.org/reports/impact-eu-iuu-regulation-seafood-trade-flows/

Orbie, J. (2011). Promoting labour standards through trade: normative power or regulatory state Europe? In Whitman, R. G. (Ed.), *Normative Power Europe* (pp. 161–184). London: Palgrave Macmillan.

Rosello, M. (2017). Cooperation and unregulated fishing: Interactions between customary international law, and the European Union IUU fishing regulation. *Marine Policy, 84,* 306–312.

Royal Ordinance on Fisheries, B.E. 2558 (2015) https://www.fisheries.go.th/law/web2/images/PR2558/6-royalfisheries.pdf

Song, A. M. (2021). Civilian at Sea: Understanding Fisheries' Entanglement with Maritime Border Security. *Geopolitics,* 1–25.

Soyer, B., Leloudas, G., & Miller, D. D. (2017). Tackling IUU fishing: Developing a holistic legal response. *Transnational Environmental Law, 7*(1), 139–163.

Steenbergen, D. J., Raubani, J., Gereva, S., Naviti, W., Arthur, C., Arudere, A., et al. (2022). Tracing innovation pathways behind fisheries co-management in Vanuatu. *Ambio, 51*(12), 2359–2375.

Stone, D. (2012). Transfer and translation of policy. *Policy Studies, 33*(6), 483–499.

Stringer, C., Burmester, B., & Michailova, S. (2022). Modern slavery and the governance of labor exploitation in the Thai fishing industry. *Journal of Cleaner Production, 371,* 133645.

Thorpe, A., Hermansen, O., Pollard, I., Isaksen, J. R., Failler, P., & Touron-Gardic, G. (2022). Unpacking the tuna traceability mosaic–EU SFPAs and the tuna value chain. *Marine Policy, 139,* 105037.

Vandergeest, P., & Marschke, M. (2020). Modern slavery and freedom: Exploring contradictions through labour scandals in the Thai fisheries. *Antipode, 52*(1), 291–315.

Vandergeest, P., & Marschke, M. (2021). Beyond slavery scandals: Explaining working conditions among fish workers in Taiwan and Thailand. *Marine Policy, 132,* 104685.

USDA Foreign Agricultural Service (2018). GAIN Report. Thailand. https://apps.fas.usda.gov/newgainapi/api/report/downloadreportbyfilename?filename=Seafood%20Report_Bangkok_Thailand_5-8-2018.pdf

Wilhelm, M., Kadfak, A., Bhakoo, V., & Skattang, K. (2020). Private governance of human and labor rights in seafood supply chains: The case of the modern slavery crisis in Thailand. *Marine Policy, 115,* 103833.

Wongrak, G., Hur, N., Pyo, I., & Kim, J. (2021). The Impact of the EU IUU Regulation on the Sustainability of the Thai Fishing Industry. *Sustainability, 13*(12), 6814.

Yea, S. (2022). Human trafficking and jurisdictional exceptionalism in the global fishing industry: a case study of singapore. *Geopolitics, 27*(1), 238–259.

Yea, S., Stringer, C., & Palmer, W. (2022). Funnels of Unfreedom: Time-Spaces of Recruitment and (Im) Mobility in the Trajectories of Trafficked Migrant Fishers. *Annals of the American Association of Geographers,* 1–16.

3 Can Anti-IUU Trade Measures Diffuse to Other Market Countries? Case Study of Australia

Kate Barclay[1]

Introduction

The international trade regime's acceptance of unilateral trade measures by the EU, US and Japan against IUU fishing has opened up potential pathways for policy diffusion – for more countries to adopt similar measures for their seafood imports. If trade-related measures on seafood imports spread, further reducing the markets that will allow entry to product not documented as being legally caught, the effects on seafood production and supply chains could be profound.

In this chapter we look at Australia. Australia is not a big seafood market, with a population of a little over 25 million and relatively low per capita annual seafood consumption of less than 15kg (Department of Agriculture, Fisheries and Forestry, 2022). Australia, however, has been an active participant internationally in creating catch documentation schemes to prevent the trading of illegal, unreported or unregulated (IUU) catch. It has been a strong proponent of preventing overfishing, is closely aligned with the US, EU and Japan in international relations, and imports well over half of its seafood. For these reasons, Australia could be expected to be the kind of state to which anti-IUU trade measures could diffuse, in terms of Australia emulating the EU and other seafood importing authorities in implementing anti-IUU trade measures on imports.

Both Australia and Thailand are potential 'receivers' of policy diffusion from the EU, US and Japan, but in different ways. Policy diffusion by emulation is a different kind of diffusion than the coercive type of diffusion considered in the case of Thailand (chapter 2). Thailand's diffusion is through being on the receiving end of anti-IUU measures on its seafood exports, and being forced to implement the policy in order to be able to continue exporting to the EU. Australia is already compliant with anti-IUU measures for its exports, so the question we consider in this chapter is whether Australia would implement anti-IUU trade measures on imports to its markets, with the only impetus being whether Australian decision makers decide an anti-IUU trade measure will be

DOI: 10.4324/9781003296379-3

useful in preventing IUU, and to 'level the playing field' between imports and the heavily regulated domestic seafood industry.

In this case we look at the internal factors affecting policy diffusion on the receiving end. Scholars of policy diffusion have pointed to the need to consider how policy ideas are accepted, rather than focussing only on the political relevance of the idea itself, as part of considering the conditions that must be generated before policy change is possible (Stone, 2012, p. 489). Legacies of existing administrative structures and domestic policy discourses are some of these conditions (Stone, 2012, p. 485; Steenbergen, Song, & Andrew, 2022). We find that domestic administrative structures and discourses about what kinds of regulation are appropriate at different points along seafood supply chains have acted as obstacles preventing Australia from emulating the anti-IUU trade measures (Garcia Garcia, Barclay & Nicholls, 2021). The generally positive conditions for diffusion noted earlier have hitherto not been able to overcome these contextual obstacles. A new Labor government, which came to power at the federal level in 2022, has indicated it may be more willing than the previous Liberal National Party to consider anti-IUU trade measures, but at the time of writing, Australia still did not have anti-IUU trade measures in place.

The chapter first details the methods and data used in this research on Australian seafood policy. The chapter then examines the historical background of the construction of IUU as a policy object in the Australian context. We go into further depth on the objectives of Australian fisheries management and the boundaries drawn between policy areas – sustainability of fisheries, trade and the regulation of food. Deliberation on the related policy areas of Country of Origin Labelling (CoOL) and standardisation of the naming of seafood at the point of sale provide rich material for considering policy positions regarding anti-IUU trade measures. Assessing the domestic context in terms of enabling or preventing policy diffusion is useful to chart the potential evolution of trade-related measures against IUU fishing and, more generally, potential pathways for greater compatibility between environmental provisions and multilateral trade regulations.

Methods and data

The data for this chapter was collected by Sonia Garcia Garcia for her doctoral research and includes interviews and observations as well as policy texts as data sources starting in September 2017 and ending August 2019 (Leipold et al., 2019, p. 449, see Table 3.1).

Interviewees were broadly categorised into government (fisheries managers, environmental managers, policy officers), research (researchers inside and outside academia and research providers), industry (fishers, aquaculturalists, seafood producer group representatives, retailers, wholesalers,

Table 3.1 Research data

Types of data	Subtype	Data-collection method
Observation	Event ethnography	Note-taking of paper presentations, discussions and event documentation for Seafood Directions conference, Sydney, September 2017
Documents	Research (literature review) Policy documents	Theoretical and snowball sampling Use of 'fields of action' classification (Wodak, 2001, p. 68)
Interviews	38 (face-to-face, in-depth, semi-structured, mostly individual, 3 groups) 30 men, 11 women	Audio recording and transcripts Purposive and snowball recruitment (Maxwell, 2013, pp. 89–91)

Adapted from Creswell (2003, p. 186).

restauranteurs and consultants) and civil society actors (consumer group representatives and environmental NGO campaigners). A number of participants had overlapping roles. The total number of interviews was 38, 35 of them one-to-one, and three with two interviewees from the same organisation. Interviews were semi-structured, with a common block of questions on the activity of the respondent, their definitions of sustainability, and the main issues in labelling and traceability, followed by questioned tailored to their role(s).

Documents were compiled according to Wodak's (2001) 'fields of action' classification (Table 3.2). From these, one particular document contained a wealth of material about IUU: a 2014 parliamentary inquiry in the Australian Senate, *Current requirements for labelling of seafood and seafood products* (Commonwealth of Australia, 2014). The inquiry provided an extensive public record of stakeholder discussions around regulation of seafood imports, including IUU fishing. The inquiry documentation includes records of 2 hearings, 25 submissions and 1 report, as well as speeches and media releases produced during the inquiry. The inquiry material was coded in NVivo together with the interview transcripts.

The policy construction of IUU fishing by the Australian Government

Australia is a minor player in global seafood terms, with around 1% of the global trade value, but it is firmly established in seafood trade networks of the Pacific region. Its fisheries are oriented towards regional exports of high value seafood to China, Vietnam, Japan and Hong Kong (Steven, Mobsby & Curtotti, 2020, p. 25). The export of products such as

Table 3.2 Categorisation of documents

Legislative instruments	Regulatory procedures	Executive and administration	Communication	Political control
Acts and regulations (11) International agreements and conventions (4) Resolutions (2)	Consultation (working groups, meetings) (1) Research and government reports and related documents (12)	Policy papers (3) Government positions (3) Strategic plans (6) Guidelines (3)	Press releases (1) Speeches (1) Factsheets (3) Media materials (interviews, documentaries, news, leaflets) (39)	Parliamentary inquiries and related docs (3 inquiries)

Adapted from Wodak (2001, p. 68).

abalone, bluefin tuna and rock lobster accounted for a gross value of production (GVP) of AUD 1.58 billion out of an overall GVP for fisheries and aquaculture of AUD 3.58 billion in 2019 (Steven, Mobsby & Curtotti, 2020, pp. 2, 5), before exports markets, especially to China, were disrupted during the COVID-19 pandemic. Imported seafood, mainly from Thailand, China, Vietnam and New Zealand, makes up over 60% of the overall seafood consumption by volume (Steven, Mobsby & Curtotti, 2020, pp. 25, 36; Department of Agriculture, Fisheries and Forestry, 2022).

Historically, Australia was a key player in the construction of IUU fishing of Patagonian toothfish (also known as Chilean sea bass) as an international issue within the CCAMLR (Österblom & Sumaila, 2011). Australia promoted holistic approaches to fight IUU fishing internationally and engaged actively in drafting measures such as the catch documentation scheme for toothfish in the CCAMLR (Agnew, 2000, p. 367); and the trade information scheme for Southern Bluefin Tuna in the Convention for the Conservation of Southern Bluefin Tuna (CCSBT) (Department of Agriculture, Fisheries and Forestry, 2005, p. 37). Australia also intervened actively in the drafting and negotiation of international instruments such as the International Plan of Action to Prevent, Deter and Eliminate Illegal, Unreported and Unregulated Fishing (IPOA-IUU), the Port State Measures Agreement (PSMA) and the Regional Plan of Action to Promote Responsible Fishing Practices including Combating Illegal, Unreported and Unregulated Fishing (RPOA-IUU). Finally, Australia participated in working groups on IUU such as the High Seas Task Force in the UN and in the OECD (Department of Agriculture, Fisheries and Forestry, 2005, p. 5; Department of Agriculture, 2014, p. iii).

Australia's efforts to address IUU fishing were framed from the beginning as a means to protect national resources:

> We were having significant problems in the sub-Antarctic with Patagonian toothfish poaching and that's where this whole process of IUU came from because not only we couldn't identify the owners of the vessels, we couldn't prosecute anybody, we couldn't follow any trade and we were genuinely annoyed and we went to the FAO the first time and the FAO told us to read the compliance manual. We said thanks for your assistance, we've already read that we want to do something a bit more. (Interview respondent, fisheries manager)

The two National Plans of Action against IUU fishing, published in 2005 and 2014 (Department of Agriculture, Fisheries and Forestry, 2005; Department of Agriculture, 2014), show an evolution in the approach to the prevention of IUU fishing from a militarised approach to cooperative action and regional cooperation in the Pacific, and acknowledge the economic and social impacts of IUU fishing (Department of Agriculture, 2014, p. 2). However, this evolution did not involve moving towards using trade measures on imports to Australian markets, as happened in the EU. The Australian plans rested on the assumption that monitoring, control and surveillance measures on fishing and a careful port policy regarding distant water fishing for overseas markets are sufficient to block access of unlawful products to Australia's domestic market, for two reasons:

> Given the very small quantities of fish which foreign fishing vessel operators have sought to land in Australia, the actual market-related implications have to date been insignificant.
>
> (Department of Agriculture, Fisheries and Forestry, 2005, p. 36)

> Given the limited extent of IUU fishing involving Australian-based operators – other than in the mainly criminal activities of domestic groups involved in illegal abalone and rock lobster fishing and trafficking discussed elsewhere in the AUS-NPOA-IUU – there has been little need to date to respond in Australia to the provisions of IPOA paragraphs 73 and 74, which call for action against importers, trans-shippers, buyers, consumers, bankers and others who may do business with IUU fishers or engage in activities that support IUU fishing.
>
> (Department of Agriculture, Fisheries and Forestry, 2005, p. 38)

The section of the Australian National Plan of Action (Department of Agriculture, 2014) on market-related measures in the IPOA-IUU acknowledged the discussion of traceability underway in the international sphere but did not advance it in the Australian context. Rather, it situated trade-related measures as subsidiary to fisheries management measures; it situated traceability under the jurisdiction of the *Australia New Zealand Food Standards Code* and circumscribed Australian participation in catch documentation schemes to two Regional Fisheries Management Organisation catch documentation schemes, CCAMLR and CCSBT (Department of Agriculture, 2014, p. 9), both of which are enforced through import measures in other countries. The role of traceability in preventing IUU-sourced products entering domestic markets mentioned in the US and EU IPOA-IUU documents – transparency, prevention of fraud, level playing field – is absent from the 2014 Australian document and the potential use of traceability to prevent IUU-sourced products entering the Australian market is left unsaid. This construction of IUU as a transnational crime connected to overseas markets largely summarises Australia's policy approach:

The market is of interest to us but usually other agencies will deal with that. We work with Interpol. We work with Sea Shepherd. We work with a whole bunch of groups of people to prevent and stop IUU fishing. Now, inevitably that does involve markets because you need market intelligence to know where this product is going so you can track it. Groups like Interpol and Sea Shepherd and others who have their own networks of people around the world are very useful in that regard. That helps the operational side know where these boats are likely to be, where they're going to be pulling into port to offload fish and transport it through the supply chains. We get all that intelligence that comes back to us and then we can deploy our staff and the assets we have in the right places at the right time. (Interviewee, fisheries manager)

Whether or not IUU-sourced products are actually entering Australian markets, however, remains to be explored. The literature review and interviews conducted for this research uncovered no studies investigating the legality of fishing of seafood imported to Australia. Two studies on seafood mislabelling have been conducted: a pilot survey conducted in 2003 (Food Standards Australia New Zealand, 2003) and a study in Tasmania in 2015 (Lamendin, Miller, & Ward, 2015). These present an inconclusive picture, the first one finding 23% of mislabelling in two species sampled across the country, and the second finding inaccuracies in the labelling of 38 samples from 15 fishmongers in Tasmania (Lamendin, Miller, & Ward, 2015, p. 438, 442). The Australian government's approach to IUU fishing as restricted to strategic interests and the protection of valuable Australian fishery exports, and not relevant for imported seafood entering domestic markets has, therefore, not been an evidence-based approach.

Boundaries and objectives of fisheries management in Australia

The disconnect between strong regulation to prevent overfishing in domestic commercial fisheries while refusing to regulate for sustainability in seafood markets reflects the boundaries of fisheries management and its objectives in Australian policy over the past couple of decades.

The foundational boundary is that of IUU being something that is regulated at the fishing/harvesting node of seafood supply chains. This construction is not unique to Australia and is a prevalent framing in the field of fisheries management. For example, an IUU fishing Index published in 2019 (Macfadyen et al., 2019) measures the 'degree to which *coastal* states are exposed to and effectively combat IUU fishing' (p. 1, emphasis added). In that index, Australia scored as having extremely low levels of IUU, ranking 138th out of 152 countries. The disconnect comes in leaving IUU only at the fishing node, when Australia, as noted earlier, was centrally involved in setting up some of the international catch documentation schemes that enabled regulation further along supply chains.

Another policy boundary at play in Australian approaches to IUU is that between federal and state/territory jurisdictions. The prevention of IUU fishing in international fisheries in which Australian fleets are involved is a policy objective at the federal level. The Australian Fisheries Management Authority (AFMA) manages all fisheries beyond three nautical miles from the low water mark including key fisheries signalled as the main target of IUU fishing, like tuna and toothfish. International fisheries negotiations are handled by the federal Department of Agriculture, Fisheries and Forestry (DAFF), with AFMA. Much of Australia's commercial fisheries, however, are coastal and under the jurisdiction of the states. The states have a strong mandate to prevent overfishing in commercial fisheries and target illegal commercial fishing as part of that, but do not address 'IUU' as it is constructed in the international sphere. Australian fisheries destined for export markets are cleared by federal agencies, including as legal for international catch documentation schemes where necessary. There is regulation of domestic fisheries supply chains to try to prevent illegally caught fish from entering markets, and there is food safety regulation, both of which are admininstered under state and territory jurisdiction. Seafood imports, regulated only for food safety, come under the independent, supranational authority Food Standards Australia New Zealand (FSANZ).

The disconnect is also related to the strong primacy of preventing overfishing as an objective for Australian fisheries management, and prioritising recreational fishing over commercial fishing, at federal and state/territory levels, and the relatively weak policy objective for the maintenance or development of domestic commercial fisheries. Australia has integrated the principles of Ecologically Sustainable Development (ESD) across all jurisdictions—federal and in each of the states and territories— and Ecosystem Based Fisheries Management (EBFM) has been adopted in a number of jurisdictions (Pascoe et al., 2019, p. 644). However, operational objectives have focused on the biological component of sustainability, with much less clarity on the economic (Emery et al., 2017) and only recent attention to the social (Barclay, 2012; Pascoe et al., 2019, p. 645). Studies on broader economic benefits than profitability of commercial fishing are only just emerging (Abernethy et al., 2020; Voyer et al., 2016), prompted not by government agencies but by industry bodies needing to demonstrate the contributions of commercial fishing to regional economies and the national economy in order to improve the industries' public image and position in policy negotiations (Fisheries Research and Development Corporation, 2020). A key impetus behind anti-IUU trade measures in the EU has been to 'even up the playing field' between domestic fisheries and importing fisheries, since domestic fisheries shoulder regulatory costs for sustainability, whereas importing fisheries may not. Australian governments have, since the 1990s, caused domestic commercial fishing industries to shrink and pushed them out of fishing grounds, in order to prevent overfishing or to prioritise recreational fishing over commercial fishing (for details of such policy trajectories in the states of New South Wales and Victoria see Abernethy et al. (2020); Barclay et al. (2020); Minnegal and Dwyer (2008). In this policy context it has been unlikely that government would introduce anti-IUU measures to protect the viability of domestic commercial fishing industries.

Another boundary within Australian fisheries management that makes adoption of anti-IUU trade measures less likely is that the regulatory pursuit of biological sustainability in the management of fisheries is seen as belonging mainly or wholly in the harvest space, and not along whole supply chains. The regulation of seafood downstream as it heads towards consumers is placed under the food regulatory system. The Australian Consumer Law and food standards issued by Food Standards Australia New Zealand (FSANZ) regulate the conditions to be fulfilled as food passes along the supply chain. Food policy responsibilities fall into the Health portfolio of the Commonwealth and the States, and enforcement corresponds to the Australian Competition and Consumer Commission and the food authorities in the states. The sustainability of food production methods are considered to be a consumer value that is left to voluntary, industry-driven initiatives rather than being regulated by government. The lack of sustainability objectives for the management of commercial fisheries in terms of their broad economic or social sustainability and the framing of IUU fishing as an issue only affecting exports means that regulatory oversight of seafood beyond the point of harvest is tied to agencies with no responsibility for the sustainability of fish stocks, Australian or otherwise.

This boundary is visible in the way traceability is used in seafood in Australian markets. The concept of 'traceability' in food was initially for food safety, so as to be able to track down all contaminated food for product recalls. Traceability in seafood has expanded out from food safety and underpins catch documentation schemes for anti-IUU measures as well as sustainability requirements in certification schemes – so as to be able to claim at the consumer end of the chain that the food was fished as it should have been. To date in Australian policy regarding seafood traceability has firmly remained a food safety mechanism only.

In the context of this structure of policy boundaries and objectives, demands for improving sustainability requirements for all seafood consumed in Australia falls 'between the cracks' of the Australian regulatory structure, lacking support from the public administrators responsible for the management of fisheries resources. As summarised by a participant,

> That's a fact that there were no Australian government requirements around the sustainability of any incoming seafood. There was no requirement there for that to be looked at or addressed by anybody. Whereas in the EU there are rules now, America's just brought in rules along that kind of lines and [here it] is not on anybody's radar. Biosecurity is on one branch of DOA's [the Department of Agriculture's] radar, food safety is another branch of DOA's radar, CITES species are supposedly on the radar of the Department of Environment but I don't think anyone's actually checking. (Interviewee, consultant)

The policy tussle over Country of Origin Labelling for seafood in food service outlets (takeaway food shops, restaurants, cafes, etc) shows that the federal government and most of the state governments have clearly refused to address fisheries management concerns at the retail end of seafood supply chains. Here we see another disconnect – this one between (1) the domestic fishing industry which wants government regulation of seafood labelling to avoid fraud and so customers are aware of where their seafood comes from; (2) fisheries managers who construct their responsibility regarding sustainability as being for Australian commercial fisheries and only in the harvest node of supply chains; and (3) food system regulators responsible for labelling and traceability, who are only concerned with food safety.

If anti-IUU measures on seafood imports implemented in the EU, US and Japan are to diffuse to Australia, Australian fisheries management agencies will have to become involved downstream from the harvest node, and also to develop Australian traceability requirements for the legality of overseas fisheries, such as catch documentation schemes. For that reason the Country of Origin Labelling case is an interesting one to help understand the willingness and bureaucratic structural impediments to fisheries management being conducted at the market end of supply chains, and using traceability as a tool for legality of catches in the Australian context.

Fishing industry and marine conservation advocates launched campaigns in the 2010s to try to have government mandate that the country of origin of seafood be shown clearly at the point of sale. Over 80% of consumers surveyed in NSW and Victoria have said they would prefer to buy Australian seafood over imports, but around a third report that they do not know where their seafood is from (Abernethy et al., 2020; Voyer et al., 2016). Since well over half of all seafood consumed in Australia is imported, if the country of origin is clearer to consumers, it seems feasible purchasing habits could shift.

Labelling requirements and traceability came under the scrutiny of the Australian Senate in an inquiry conducted in 2014 on the requirements for the labelling of seafood (Commonwealth of Australia, 2014). The issue had been publicized at that time through an Australian Broadcasting Corporation (ABC) television series called 'What's the Catch' about sustainability problems in the seafood consumed in Australia, which argued for sustainability regulation for imports, and a concurrent Greenpeace campaign on the same topic. The inquiry debated labelling and traceability requirements in the context of two demands: one, to make use of the existing Australian Fish Names Standard mandatory, a possibility enabled in the food standard regulatory framework pushed by a broad coalition of industry actors and environmental organisations to avoid fraud through the use of misleading names. The other demand was that food service outlets such as restaurants or fish and chip shops should specify the country of origin of their product or, at least, indicate whether it is imported.

> Some in the fishing industry were calling for better labelling, some sectors were calling for mandatory fish name standard. Others were calling for voluntary fish names standard, but for the standard to be refined more. The fishing industry was dead against some of the labelling things that we wanted included, in particular the type of fishing gear that was being used. They wanted country of origin labelling, but really, they just wanted to distinguish between something caught outside of Australia and something caught in Australia. And I am sure that that wasn't a universal, they didn't universally want that because many of them have businesses that are partly Australian production and partly importing. (Interviewee, NGO representative)

The NGO representative is alluding to the fact that not all seafood industry players saw Country of Origin Labelling as being in their best interest. The Australian fishing industry was fairly consistent, but many seafood wholesalers and retailers had interests mixed up between domestic and imported seafood, or some focussed mainly on imported seafood. Imported seafood is often cheaper, processing costs in particular are much lower in places like Vietnam and China than Australia, and often it is easier to obtain consistent volumes through imported sources. For the hospitality industry, being able to get the same kind of frozen boxes of fillets all year round from importers is cheaper and easier than sourcing fresh seafood locally, with seasonal and weather-dependent fluctuations, and cooks may have to do the filleting or other preparation themselves. To use only seafood sourced from domestic commercial fisheries means cooks and customers must be willing to embrace a range of different species, rather than having exactly the same thing on the menu all year round (Abernethy et al., 2020). Moreover, some seafood retailers have benefited from the lack of clarity about country of origin in labelling. For example, fish and chip shops in coastal locations sell cheaper imported seafood to holiday makers who assume the catch is local, or fish shops showcase shiny fresh local catch in their window, and shoppers attracted to the window display assume that the marinara mix or calamari rings are also local (Abernethy et al., 2020).

The long-standing demand of the domestic fishing industry for Country of Origin Labelling (CoOL) in all food outlets had been granted in the jurisdiction of the Northern Territory, through fisheries legislation, not through the food regulation system. The fishing industry wants CoOL for several reasons. One is as a means to establish a level playing field for the domestic produce subject to high production costs, including those regulatory costs associated with sustainable fisheries management. CoOL and the mandatory specification of standard names would also contribute to the prevention of mislabelling practices and to greater efficiency on border controls for imports. The example offered was the use of the term 'flake' for different shark species, both domestic gummy shark subject to strict management controls and five different shark species from overseas fisheries without management strategies (Commonwealth of Australia, 2014, Submission 13, pp. 2–3). Indicating country of origin of seafood in food outlets like fish and chips shops would also enable domestic producers to reap potential price premiums associated with stated consumer preferences for national produce (Lawley, 2015, p. iii).

Several aspects of the inquiry show the effects of the perceived boundaries of fisheries management on the possible roles of traceability and labelling requirements to address IUU fishing and the conditions governing seafood markets. The first one, the Senate inquiry, received 25 submissions,[2] of which four were from agencies responsible for fisheries, three were from the state governments of Northern Territory, Queensland, New South Wales and one was from the federal Department of Agriculture. Fisheries management agencies for half of the jurisdictions did not state any position in the form of a submission, which may be interpreted as saying that the requirements for seafood labelling and traceability are irrelevant to the management of fisheries in Australia. Secondly, only the joint submission by the New South Wales Food Authority and Fisheries New South Wales recommended the adoption of the Australian Fish Names Standard, a demand that had been backed by submissions from actors as diverse as fishing industry bodies, importers, fish markets, large retailers, environmental NGOs and researchers (Submissions 1, 6, 9, 10, 13, 17, 19, 20 and 24). Indeed, two submissions by fisheries agencies provided arguments to oppose the adoption of the standard on the grounds that although they had adopted its use, it still needed improvements. Thirdly, fisheries agencies submissions asserted that within food regulation sustainability is a consumer value best left to market forces (Submission 19, pp. 3-4) and consistently opposed changes to labelling or traceability that would increase regulatory burden (Submission 4, p. 1, Submission 19, p. 5, Submission 11, p. 5). Several public agencies also pointed out that trade measures could be seen as trade restrictive, and raised alerts about the potential costs involved in a regulatory process, starting with those involved in conducting consultations and cost-benefit analysis. The federal Department of Agriculture made the following statement about international developments in traceability:

> Traceability and labelling is [*sic*] attracting increasing attention in international fisheries management. Some countries are seeking more information on where and how seafood was caught and whether it is consistent with international, regional and domestic fisheries regulations. Unilateral market measures taken by an importing country can be trade restrictive in that they do not necessarily recognise equivalent or better arrangements put in place by other countries with differing approaches. Some, including the EU and the US, have already implemented market state certification requirements that have caused additional requirements for some Australian seafood exporters.
>
> (Commonwealth of Australia, 2014, Submission 11, p. 4)

The Australian federal Department of Agriculture, Forestry and Fisheries (DAFF) is responsible for the prevention of IUU internationally, but in its submission on the Fish Names Standard within Australia omitted the rationale behind unilateral trade measures, thereby dissociating the prevention of IUU fishing from traceability and labelling. State fisheries agencies, not involved in the global fora on IUU fishing or in the management of fisheries post-harvest, ignored the connection brought up by environmental NGOs and producers between traceability and lawful sourcing of seafood products that is now an accepted strategy internationally and which other parts of the Australian government were active in establishing. Seeing their policy mandate as being to sustain domestic fish stocks, the state fisheries management agencies deferred traceability and labelling of imported seafood to the food regulatory framework – disconnecting it from the potential impact on domestic fisheries of IUU fishing in overseas fisheries competing in Australian markets. The lessons learned by Australian agencies at the international level in the prevention of IUU fishing were not brought home, and the state agencies ignored that the choice of policy tools to address sustainability in the post-harvest space has direct implications for those public bodies with a regulatory responsibility to pursue sustainability in domestic fisheries.

Unsurprisingly, the 2014 inquiry did not result in mandated use of consistent naming in the labelling of seafood and Country of Original Labelling was only partially applied. In 2016 new legislation made Country of Origin Labelling mandatory for retail outlets selling raw or packaged seafood (fish shops and supermarkets), but left it optional for the food service sector selling prepared seafood dishes (fish and chip shops, restaurants, takeaway shops, etc.). Over 40% of overall seafood sales in Australia are from the food service sector (Productivity Commission, 2016, p. 270), so this ommision was significant. In 2020 a review of the effects of the implementation of the 2016 Country of Origin Labelling regulations was conducted. By this stage, groups representing wholesalers who imported seafood as well as fishers were united in calling for mandatory CoOL to be extended to food service, noting that the 2018 introduction of mandatory CoOL for fish shops had not caused major problems for industry (Sydney Fish Market, 2020; Seafood Industry Australia, 2020). The evaluation found, however, that the food service sector had been hard hit by COVID-19 responses so additional regulatory burden in the form of CoOL should not be applied (Deloitte Access Economics, 2021). The federal government accepted this finding and stated in February 2022 that the CoOL regulations would not be changed (Department of Industry, Science and Resources, 2022).

The CoOL policy discussion shows that in the recent past Australian fisheries management agencies, including at the federal level where responsibility for any anti-IUU measures would sit, have promoted a structure of policy objectives and boundaries that is not amenable to anti-IUU trade measures being applied to seafood imports. This structure limits fisheries management responsibilities to the fishing node of supply chains, and limits government involvement in post-harvest nodes of the chain to biosecurity and food safety. Efforts to ensure sustainability along supply chains using traceability techniques are constructed as a consumer choice issue rather than a government responsibility. Spokespeople for fisheries management agencies have called measures other than biosecurity or food safety applied to seafood imports potentially trade-restrictive (and therefore a bad thing).

However, in May 2022 there was a change of federal government, with the centre-right Liberal National Party government that had been in power since 2013 replaced by a centre-left Labor government. It appears that Labor politicians may be more amenable to the kinds of government regulation that could enable the diffusion of the anti-IUU trade policy. The new Minister for Agriculture, Fisheries and Forestry, Murray Watt, stated in his speech to open the seafood industry conference Seafood Directions (13–15 September 2022, Brisbane) that he would make Country of Original Labelling mandatory for food service outlets. Left-leaning policy think-tank the McKell Institute hosted a meeting about IUU fishing, with former Labor Minister for Trade Craig Emerson as the main speaker in November 2022 to initiate policy guidance and research that can support Australia on creating an IUU fishing policy that applies to seafood imports. So although the policy environment to date has not been fertile ground for anti-IUU trade measures to diffuse to Australia, that may change in the near future, and diffusion might become possible.

Another point worth noting regarding the Australian case, in light of the discussion of application of the EU IUU Regulation in the Thai case, is that the labour and human rights conditions of production of seafood imported to Australia have been largely missing from fisheries policy discussion to date. There are general policy discussions about modern slavery and about government and corporate responsibility to ensure workforces' human rights are protected in the making of products sold in Australian markets. The *Modern Slavery Act 2018* requires all large companies to annually report on risks of modern slavery in their operations and supply chains, including overseas suppliers. A report on the first two years of implementation of this Act finds that two out of three companies covered by the Act are still failing to properly report slavery risks (Dimshaw et al., 2022). The report highlights seafood processing in Thailand as one industry with a high risk of slavery, but labour abuses in imports is not high on the agenda in Australian seafood policy circles. There has been more discussion on sustainability of fisheries, in terms of preventing biological overfishing, and the disconnect between regulating to make domestic commercial fisheries sustainable and not regulating overseas commercial fisheries supplying Australian markets.

Conclusion

Sonia Garcia Garcia has argued that Australia's overall policy stance towards fishing and seafood is characterised by disconnections (Garcia Garcia, Barclay & Nicholls, 2020). One key disconnect is that despite Australia's historical and continuing role in promoting trade-based measures to prevent IUU fishing for other markets, and the importance of international trade for its own seafood sector, Australian authorities have not implemented anti-IUU measures for its own markets. Australian regulation of imports has been limited to food safety and biosecurity. This relates to a disconnect in the domestic policy context between policy towards commercial fishing in Australia – in which sustainability of fish as a natural resource is the most prominent objective – and policy towards fish as food downstream in value chains – in which food safety is the most prominent objective. Australian policy towards fish as food in markets relegates 'sustainability' to a consumer choice and private sector concern, not as something for government regulation.

The Australian case shows that even when some factors are in place for policy diffusion – such as a strong history of participation in international action against IUU, including a history of aligned action with those states on anti-IUU trade measures for other markets, and close international relations with the anti-IUU trade measure policy-initiating states – it may not occur due to domestic factors confounding the emulation. Existing domestic bureaucratic structures and policy objectives for fish as natural resource vs fish as food with health and biosecurity implications have thus far acted against Australia emulating the EU, US and Japan in implementing anti-IUU measures on seafood imports. The Australian case thus demonstrates some of the ways domestic policy contexts may influence policy diffusion.

Notes

1 The empirical work used in this chapter is adapted from the doctoral research of Sonia Garcia Garcia (2019; Garcia, Barclay & Nicholls 2021) for which Kate Barclay was primary supervisor. For personal reasons, Dr. Garcia voluntarily withdrew from the writing of this book, giving permission to Kate Barclay to publish the research.
2 The list of submissions and the documents are available at https://www.aph.gov. au/Parliamentary_Business/Committees/Senate/Rural_and_Regional_Affairs_ and_Transport/Seafood_labelling/Submissions (retrieved February 1, 2023).

References

Abernethy, K., Barclay, K., McIlgorm, A., Gilmour, P., McClean, N., & Davey, J. (2020). Victoria's fisheries and aquaculture: Economic and social contributions. Fisheries Research and Development Corporation (FRDC 2017-092), University of Technology Sydney.

Agnew, D. J. (2000). The illegal and unregulated fishery for toothfish in the Southern Ocean, and the CCAMLR catch documentation scheme. *Marine Policy*, *24*(5), 361–374. doi:10.1016/S0308-597X(00)00012-9

Barclay, K. (2012). The social in assessing for sustainability: Fisheries in Australia. *Cosmopolitan Civil Societies: An Interdisciplinary Journal*, *4*(3), 38–53. doi:10.5130/ccs.v4i3.2655

Barclay, K., Davila, F., Kim, Y., McClean, N., & McIlgorm, A. (2020). Economic analysis & social and economic monitoring following the NSW Commercial Fisheries Business Adjustment Program. Report commissioned by the New South Wales (NSW) Department of Primary Industries. Retrieved January 31, 2023 from, https://www.dpi.nsw.gov.au/__data/assets/pdf_file/0007/1256128/

Economic-analysis-and-Social-and-Economic-monitoring-following-the-NSW-Commercial-Fisheries-Business-Adjustment-Program.pdf

Commonwealth of Australia (2014). *Current requirements for labelling of seafood and seafood products.* Canberra: Parliament of Australia. Retrieved February 27, 2023 from, https://www.aph.gov.au/Parliamentary_Business/Committees/Senate/Rural_and_Regional_Affairs_and_Transport/Seafood_labelling

Creswell, J. W. (2003). *Research design: Qualitative, quantitative, and mixed methods design.* London: SAGE.

Deloitte Access Economics (2021). Evaluation of Country of Origin Labelling reforms. Report prepared for Department of Industry, Science, Energy and Resources. Deloitte Access Economics, Melbourne. Retrieved February 24, 2023, from, https://www.industry.gov.au/publications/country-origin-labelling-food-reforms-evaluation

Department of Agriculture (2014). *Australia's second national plan of action to prevent, deter and eliminate illegal, unreported and unregulated fishing,* Canberra: Department of Agriculture. Retrieved February 27, 2023 from, https://www.agriculture.gov.au/sites/default/files/sitecollectiondocuments/fisheries/iuu/aus-second-npoa-iuu-fishing.pdf

Department of Agriculture, Fisheries and Forestry (2022). Fisheries and aquaculture statistics website, ABARES, Canberra. Retrieved January 31, 2023 from, https://www.agriculture.gov.au/abares/research-topics/fisheries/fisheries-and-aquaculture-statistics

Department of Agriculture, Fisheries and Forestry (2005). *Australian national plan of action to prevent, deter and eliminate illegal, unreported and unregulated fishing.* Canberra: Department of Agriculture, Fisheries and Forestry. Retrieved February 27, 2023 from, *Australian national plan of action to prevent, deter and eliminate illegal, unreported and unregulated fishing*

Department of Industry, Science and Resources (2022). Country of Origin Labelling evaluation report released. Media release 25 February. Retrieved January 31, 2023 from, https://www.industry.gov.au/news/country-origin-labelling-evaluation-report-released

Dimshaw, F., Nolan, J., Hill, C., Sinclair, A., Marshall, S., McGaughey, et al. (2022). Broken promises: Two years of corporate reporting under Australia's Modern Slavery Act. Melbourne: Human Rights Law Centre. Retrieved February 24, 2023 from, https://www.hrlc.org.au/reports/broken-promises

Emery, T. J., Gardner, C., Hartmann, K., & Cartwright, I. (2017). Incorporating economics into fisheries management frameworks in Australia, *Marine Policy, 77,* 136–143. doi:10.1016/j.marpol.2016.12.018

Fisheries Research and Development Corporation (2020). *National fisheries and aquaculture industry social and economic contributions study.* FRDC project 2017-210, FRDC, Deakin West, ACT. Retrieved February 27, 2023 from, https://www.frdc.com.au/sites/default/files/products/2017-210-DLD2.pdf

Food Standards Australia New Zealand (2003). *A pilot survey on the identity of fish species as sold through food outlets in Australia,* Retrieved August 1, 2019, from, http://www.foodstandards.gov.au/publications/pages/pilotsurveyontheidentityof-fish/Default.aspx

Garcia Garcia, S. (2019). Policy disconnections in the regulation of sustainable seafood in Australia. PhD thesis, University of Technology Sydney, Australia.

Retrieved February 24, 2023 from, https://opus.lib.uts.edu.au/handle/10453/142369

Garcia Garcia, S., Barclay, K., & Nicholls, R. (2021). Can anti-illegal, unreported, and unregulated (IUU) fishing trade measures spread internationally? Case study of Australia. *Ocean & Coastal Management, 202*, 105494. 10.1016/j.ocecoaman.2020.105494

Garcia Garcia, S., Barclay, K., & Nicholls, R. (2020). The multiple meanings of fish: Policy disconnections in Australian seafood governance. In E. Probyn, K. Johnston, & N. Lees (Eds.), *Sustaining seas: Oceanic space and the politics of care* (pp. 75–86). London & New York: Rowman & Littlefield.

Lamendin, R., Miller, K., & Ward, R. D. (2015). Labelling accuracy in Tasmanian seafood: An investigation using DNA barcoding. *Food Control, 47*, 436–443. doi:10.1016/j.foodcont.2014.07.039

Lawley, M. (2015). *A final seafood omnibus: Evaluating changes in consumer attitudes and behaviours*, Project No. 2015/702, Bedford Park, SA: Seafood CRC. Retrieved February 27, 2023 from, https://www.frdc.com.au/sites/default/files/products/2015-702-DLD.pdf

Leipold, S., Feindt, P. H., Winkel, G., & Keller, R. (2019). Discourse analysis of environmental policy revisited: Traditions, trends, perspectives. *Journal of Environmental Policy & Planning: Discourse, power and environmental policy: discursive policy analysis revisited, 21*(5), 445–463. doi:10.1080/1523908X.2019.1660462

Macfadyen, G., Hosch, G., Kaysser, N., & Tagziria, L. (2019). *The IUU Fishing Index, 2019*, Retrieved November 15, 2019, from https://globalinitiative.net/wp-content/uploads/2019/02/IUU-Fishing-Index-Report-web-version.pdf

Maxwell, J. A. (2013). *Qualitative research design: An interactive approach* (3 ed.), Thousand Oaks, CA: SAGE.

Minnegal, M., & Dwyer, P. D. (2008). Managing risk, resisting management: Stability and diversity in a southern Australian fishing fleet. *Human Organization, 67*(1), 97–108. Retrieved February 24, 2019, from 10.17730/humo.67.1.x38g60k463p26855

Österblom, H., & Sumaila, U. R. (2011). Toothfish crises, actor diversity and the emergence of compliance mechanisms in the Southern Ocean. *Global Environmental Change, 21*(3), 972–982. Retrieved February 2, 2023, from, 10.1016/j.gloenvcha.2011.04.013

Pascoe, S., Cannard, T., Dowling, N. A., Dichmont, C. M., Breen, S., Roberts, T., & et al. (2019). Developing harvest strategies to achieve ecological, economic and social sustainability in multi-sector fisheries, *Sustainability, 11*(3), 644–665. doi:10.3390/su11030644

Productivity Commission (2016). *Marine fisheries and aquaculture, final report.* Canberra: Productivity Commission. Retrieved February 27, 2023 from, https://www.pc.gov.au/inquiries/completed/fisheries-aquaculture/report/fisheries-aquaculture.pdf

Seafood Industry Australia (2020). 'Aussie! Aussie! Aussie?': Seafood industry calls on Aussies to support origin labelling review. Media release 6 August. Retrieved January 31, 2023, from http://seafoodindustryaustralia.com.au/aussie-aussie-aussie-seafood-industry-calls-on-aussies-to-support-origin-labelling-review/

Steven, A. H., Mobsby, D., & Curtotti, R. (2020). *Australian fisheries and aquaculture statistics 2018* (Fisheries Research and Development Corporation project 2019-093). Canberra: ABARES. Retrieved February 27, 2023 from, https://www.frdc.com.au/project/2019-093

Stone, D. (2012). Transfer and translation of policy. *Policy Studies, 33*(6), 483–499.

Steenbergen, D. J., Song, A. M., & Andrew, N. (2022). A theory of scaling for community-based fisheries management. *Ambio, 51*(3), 666–677.

Sydney Fish Market (2020). Evaluation of country of origin labelling for food, Sydney fish market submission. Retrieved January 31, 2023, from https://www.sydneyfishmarket.com.au/Portals/0/adam/Content/qKYITYG_D0a05MqTTOLoVA/ButtonLink/200910%20Sydney%20Fish%20Market_Submission_Evaluation%20of%20Country%20of%20Origin%20Labelling.pdf

Voyer, M., Barclay, K., McIlgorm, A., & Mazur, N. (2016). *Social and economic evaluation of NSW coastal professional wild-catch fisheries: Valuing coastal fisheries*, FRDC project 2014-301, FRDC, Deakin West, ACT.

Wodak, R. (2001). The discourse-historical approach. In R. Wodak, & M. Meyer (Eds.), *Methods of critical discourse analysis* (pp. 63–94). London: SAGE.

4 Lessons from Thailand and Australia on the Diffusion of Anti-IUU Fishing Trade Policy

Alin Kadfak, Kate Barclay, and Andrew M. Song

Introduction

The EU anti-IUU regulation in force since 2010 is a significant piece of policy work aimed to improve the conservation effort of global fish stocks in light of widespread IUU fishing practices occurring around the world. Many non-EU governments have been involuntarily engaged in the process over the years with some success, and after a decade-long implementation, the EU maintains its policy drive in third countries that have trade associated with the EU. This trade-restrictive regulation exerts influence on any countries exporting seafood to the EU and creates the intended effect of reducing the occurrence of IUU fishing and improving domestic management practices in the target countries. Hence, the scope of the policy is global, outward and unilaterally driven based on the EU's market power and political 'clout'.

According to a study by Mundy (2018), the EU carding system has had significant impact on seafood trade flows from countries carded yellow and red. The majority of the countries in the study sample, including Thailand, had declined export flows to the EU around the carding announcement and the period when the EU started dialogue with those countries. However, there have been reports of significant or sudden increases in imports in some of the yellow-carded countries. These peaks represent a 'race to trade', trying to move a lot of product in anticipation of any future import ban (receiving a red card), or an offloading of products when cards are lifted and markets become available again (Mundy, 2018, p. 15).

The sanction power attached to the EU IUU regulation allows the EU to police and ban trade from countries with seafood products caught by IUU fishing practices. The EU anti-IUU regulation, however, is considered compliant with the World Trade Organisation (WTO) for two reasons. First, the trade restrictive measures are carried out before and after an official warning or sanction given to the exporting countries by the EU. Second, the EU anti-IUU fishing import blocking process is not considered discriminatory or unjustifiable because the EU has applied

DOI: 10.4324/9781003296379-4

the same standard among EU member states and third countries (Leroy, Galletti & Chaboud, 2016, p. 86). In this sense, trade measures attached to EU anti-IUU trade policy should be understood as market power used by the EU over importing countries. Acting as a voluntary agreement, 'only countries wishing to trade on the EU market need compliance' (Miller, Bush & Mol, 2014, p. 141).

This book illustrates that the EU is essentially propagating its own sustainability aims relating to IUU fishing externally, and that policy diffusion is a useful theoretical lens to explore this outward promotion of the policy in non-EU countries. The US and Japan have subsequently also established regulations against imports of seafood deemed to be IUU. Both Australia and Thailand are (potential) 'receivers' of policy diffusion from the EU, US and Japan, but in different ways. Australia may choose to implement a similar kind of policy for its own seafood imports. Thailand on the other hand, has had to adopt the EU policy under coercive conditions. In concluding this book, we pull together some ideas from the policy diffusion literature to think about our case studies to make some general observations about the policy diffusion of anti-IUU regulation on seafood imports.

First, we think about our cases in terms of the four types identified by policy diffusion scholars – learning, competition, coercion and emulation (Braun & Gilardi, 2006; Gilardi & Wasserfallen, 2019; Shipan & Volden, 2008). Neither of our cases seem to be cases of diffusion as the result of competition – possibly the US and Japan establishing their own anti-IUU trade measures after the EU did were cases of diffusion as a response to competition between similarly large seafood markets who also have significant seafood production industries. The cases of Thailand and Australia, however, seem to be coercion in the case of Thailand, and learning and/or emulation in the case of Australia. We explore what the different forms of policy diffusion in the two cases reveal about the nature and possibilities of the spread of anti-IUU trade restrictions.

Second, Gilardi and Wasserfallen (2019) point out that there is politics in policy diffusion – it is not simply a technocratic spreading of best practice. Diffusion by coercion is clearly political, with a powerful state imposing the adoption of policy on another. However, diffusion by learning – adoption based on a rational judgement about whether a policy is effective – may also be political in that is often driven by the political effects of policies, especially electoral effects, rather than only being driven by the belief that a policy is best practice. Politics also plays a role in diffusion by emulation, which Gilardi and Wasserfallen (2019) distinguish from learning (based on rational judgement) by saying that emulation is based on moral or ethical judgements about the appropriateness of a policy. Ideology plays a role in perceptions of how appropriate a policy is for emulation. Diffusion may thus vary depending on which party is in power. For example, liberal governments often adopt policies for human and minority rights, while conservative governments often adopt policies for stricter immigration control. As Gilardi and Wasserfallen (2019) point out, however, there is often overlap between learning and emulation. With ideologically driven diffusion the learning is selective, politicians 'cherry pick' the evidence for the policies they like, and ignore evidence against, and ignore evidence for policies they dislike. In our cases the diffusion of anti-IUU fishing trade policy reflects international political economy, ideological politics where domestic constituencies are key audiences for policy performances, and the mundane politics of jurisdictional turf marking within governments.

Third, we bring in the policy translation concept into the analysis of the two cases. While policy diffusion emphasizes the imitation of meaning constructed from one policy context to the political structure of the new context, Johnson and Hagström (2005) argue that there are three contributions to policy diffusion that can be made from the policy translation field. Policy translation helps us deepen and problematize the policy concept and idea, particularly in terms of examining local context, including social relations, and also by not assuming that policies are immutable as they travel, but are shaped by local actors (Mukhtarov, 2014). The situational characteristics of the receiving context are important. Johnson and Hagström (2005) draw ideas from early works of Latour (1986), to give weight to the importance of local actors, who transport the policy into the local organisation and translate them into action. Policy translation puts an emphasis on actors being involved in a continuous translation process through which society is constantly created and re-created. This means that the policy should be seen as an open, continuous process, as well as dependent on the societal distribution of power (Johnson & Hagström, 2005).

The Thailand case

Thailand's experiences of the EU's anti-IUU trade regulation fit the category of 'coercive' policy diffusion, but with additional add-on implications when exploring the policy results emerging on the ground. Coercion is the causal mechanism of policy diffusion when 'policies are introduced because powerful countries or international organisations enforce policy changes' (Gilardi & Wasserfallen, 2019, p. 1247). Often coercion refers to policy diffusion processes that move hierarchically, 'with policy imperatives emanating out or 'down' from powerful centers' (Peck, 2011, p. 787). Although 'coercion' implies top-down pressures, rather than horizontal interdependencies (Braun & Gilardi, 2006), in the case of Thailand the coercion lens is also useful for understanding how powerful actors manipulate incentives ('carrots') and disincentives ('sticks') to influence others actors to implement policy change (Braun & Gilardi, 2006; Simmons, Dobbin & Garrett, 2006). Although, as noted earlier, the EU measure is voluntary and thus may be seen as not fully coercive, in practice the prospect of losing access to the EU seafood market was a dire enough prospect that the yellow card constituted a big 'stick' deployed by the EU to encourage Thailand to adopt policy measures to eliminate IUU fishing.

In the Thailand case, the EU anti-IUU policy used the key term of 'corporate' actor as part of the carding condition. This meant that once the Thai government showed interest in working towards downgrading from yellow to green card, the Thai government was therefore willing to cooperate with the EU through government-to-government dialogue. We propose that Thailand is an instructive example to highlight how domestically driven European normative values are interpreted and being integrated into a broader EU external fisheries policy. Policy diffusion scholars remind us of the importance of the communicative function of policy when moving from one space to the next. As Johnson and Hagström (2005, p. 366) put it: 'policies ought to be seen as bearers and generators of meaning'. EU normative values, which we elaborate below, refer to environmental sustainability of fisheries management, including conservation measures, transparency and protection of labour rights through the decent work and anti-forced labour agenda, following the International Labour Organisation (ILO).

EU anti-IUU fishing trade policy generates meaning for different stakeholders within the Thai seafood industry, in ways that fabricate different responses. In this book we have not discussed much beyond the Thai government, boat owners and fish workers regarding the meaning of IUU fishing defined by the EU. However, we would like to recognise the works of other scholars, who have been exploring non-state actors' responses towards the EU anti-IUU policy and the immediate impacts of the yellow card and the add-on issue of labour rights. These works contribute to a broader view of seafood supply chains actors, particularly upstream actors, for whom the policy could become a risk object (Wilhelm et al., 2020). The way in which EU activated economic control during the yellow card period, has for global north consumers generated a meaning of 'distrust' in Thai seafood supply chains. Drawing from policy diffusion and policy translation literature, we analyse Thailand's case in three ways: power asymmetry; ideology and normative power; and domestic politics.

Power asymmetry

The coercion policy diffusion lens helps us explore two aspects of power asymmetry between Thailand and the EU. First, the international political economy of large market states forcing policy diffusion on developing exporting states is an obvious observation from our Thailand case. The EU has not given yellow or red cards to wealthy industrial countries nor key fishing nations like China, even when these are arguably engaged in IUU fishing (for discussion of China's approach to anti-IUU fishing see Song, Fabinyi & Barclay, 2022). Scholars have begun to criticise the power asymmetry between the EU and the carded countries, particularly in relation to reasons behind issuing cards and the process of government-to-government dialogue (Kadfak & Antonova, 2021; Miller et al., 2014).

The power asymmetry in the Thailand case is quite nuanced, and not simply forceful. The Thailand government has performed the role of a 'cooperative' partner to the EU. The EU has used mechanisms such as socialisation and partnership to ensure smooth policy translation in the Thai context (Kadfak & Antonova, 2021). Thailand has been portrayed as a successful example of EU anti-IUU policy implementation on the 10th anniversary of the policy on 11 December 2020 (EJF, 2022). However, the EU's anti-IUU ideas have been integrated into Thai national fisheries law and practices without the participation of all relevant Thai stakeholders. As we demonstrate in the Thailand chapter, boat owners, seafood processor companies, environmental and labour non-government organisations (NGOs) and workers have not been part of the dialogue. They have therefore not been able to provide reflexive voices as domestic policies were being formed, they have had these policies implemented upon them by the Thai government in a top-down manner. Such policy diffusion therefore was not participatory with all stakeholders, which raises questions about the sustainability of the policies due to potential legitimacy problems. Such concerns reflect a technical problem of this policy, being conducted only via state-to-state dialogue. Power asymmetry is observed not only in the relationship between the EU and Thailand governments, but also through relationships between the Thai government and domestic actors, who were excluded from the discussion table. In this way layers of politics of policy diffusion (Gilardi & Wasserfallen, 2019) are revealed. Moreover, looking through the lens of policy transfer at the Thailand case we see how the social relations of the local context affects diffusion. The top-down approach was possible because Thailand was having an authoritarian phase at the time (discussed below). This also means the policy has taken shape and is viewed in particular ways by stakeholders because of the authoritarian, top-down approach, and may ultimately undermine the adoption of anti-IUU policy in Thailand.

Moreover, the power asymmetry between the EU and Thailand has been connected to broader neoliberal economy and consumerism discourses beyond the policy regime. Ostensibly the Thai government acted rapidly upon receiving a yellow card because of the potential threat that the EU would implement full sanctions. The Thai government actions enabled seafood exports to the EU to remain high. According to Mundy (2018, p. 14), between 2005 and 2016 (prior to and during the yellow card period), Thailand had the highest import volume and value of exports to the EU compared to 11 other carded countries. However, our Thailand chapter in this book and other recent studies (Bhakoo & Meshram, 2021; Wilhelm et al., 2020) have shown that the meaning of the carding system goes beyond the direct economic threat of trade measures. According to our interviews, Thai government officers and major seafood chain companies were concerned about what the yellow card meant for the image of Thai seafood and the loss of trust from EU member states and other seafood markets, as much as they were concerned about the actual threat of a red card. Among the companies' representatives we talked to economic risk has been translated into reputational risk (Wilhelm et al., 2020). In response, these companies have put more energy on Corporate Social Responsibility (CSR) projects, often in collaboration with local Thai NGOs (Kadfak, Wilhelm & Oskarsson, 2023).

Ideology and normative power

As argued in Gilardi and Wasserfallen (2019, p. 1246), policy learning is heavily mediated by politics, and decision makers filter their policy experiences. This is since policy adoption, embedded in policy cycles in a classical sense, is not a mere technocratic act, but is a political process, where information is processed through ideological lenses (ibid, p. 1251). Studies of European policy have used the concept of normative power (Manners, 2002, 2011) to explain the pushing of ideological stances as policy diffusion. EU 'green ideology' or the 'European Green Deal' is translated into fisheries policy through Common Fisheries Policy (CFP). And while the CFP is focussed on the green behaviour of EU member states, EU anti-IUU policy and the Sustainable Fishing Partnership Agreements (SFPAs) are the two core policies translating normative values of environmental sustainability to different parts of the world (Kadfak & Antonova, 2021; Thorpe et al., 2022).

It is difficult to measure the effects of ideology and normative power. '[C]lean lines of cause and effect are invariably difficult to establish, even where power asymmetries are extreme' (Peck, 2011, p. 787). However, one can observe how such ideology and values are triggered in the public sphere of the receiving country. To have a closer look at Thai government policy experiences during the fisheries reform, one may confirm that the EU has successfully 'mediated' the ideology of anti-IUU fishing into Thai policy discourse. For instance, Thailand declared 'anti-trafficking' and 'combating IUU fishing' as national agenda items during the reform.

The country has taken up a proactive role in promoting the elimination of IUU fishing. For instance, according to Deputy Prime Minister General Prawit Wongsuwon in a speech given at the United Nations to make Illegal, Unreported and Unregulated (IUU) fishing an environmental crime (Wipatayotin, 2019):

> *Thailand is proud of its success in tackling IUU. We hope to see further international cooperation in dealing with the issue. We also want to see the United Nations not only considering cases of fishing destruction, but also treating IUU fishing as a crime for which the culprits must be punished.*

Not only has the EU praised the Thai government at international forums on their work to integrate EU anti-IUU regulations into Thailand's domestic fisheries management, the EU has provided Thailand with further support for anti-IUU fishing initiatives within South East Asia. Thailand is well positioned within ASEAN to take the lead role in adopting key EU anti-IUU policy and influence the other ASEAN member states. In 2019, with support from the EU, Thailand upheld its strategy of 'fighting [the] IUU agenda' in South East Asia by taking the lead in ASEAN IUU network (Kadfak & Linke, 2021). Thailand government authorities have provided technical support and knowledge exchange with the Vietnamese government during the current Vietnamese fisheries reform due having received a yellow card in 2017. The Thai government is helping to diffuse the EU's anti-IUU policy within the South East Asian region. With the help of Thailand the EU as policy sender is successfully creating 'common norms' whereby actors start to share similar views on which courses of action are appropriate and which are not, leading all actors to think the same way (Braun & Gilardi, 2006, p. 310).

Domestic politics in both policy-sender and -receiver states

EU domestic politics

When we discuss EU anti-IUU policy, it is important to keep in mind that the EU is not one political unit, but a collective of EU member states, some of which have more claims and influence over fisheries policy than others (Kadfak & Antonova, 2021), and indeed that non-state actors can also drive politics. We however would like to point out that the 'modern slavery' discourse has taken a central role in the initial response by the EU towards labour issues in seafood supply chains. According to our key informant interviews, international media and NGOs created strong pressure for the EU to include labour issues into the dialogue. In this way non-state actors played a key role in this case of policy diffusion, as part of the political landscape in the EU. Evidence-based NGOs have been exposing the problem of human and labour rights seafood supply chains (Kadfak et al., 2023). Furthermore, there has been a coalition of international NGOs working closely with the EU to improve transparency in seafood supply chains and promote policies to combat IUU fishing. This coalition consists of The Environmental Justice Foundation (EJF), Oceana, The Nature Conservancy (TNC), The Pew Charitable Trusts and World Wildlife Fund (WWF).[1] The relationship between EU and this coalition remains understudied, but it is clear from our observations that this group of NGOs have prioritised evidence-based reports and media outreach to create norms and activism.

One of our informants did mention that the EU has internally discussed among different European Directorates to which extent that EU anti-IUU policy should expand to include the labour issue. Until then the 'illegal' in IUU referred to breaking fisheries laws, not labour laws. If the EU incorporates a labour rights agenda into formal anti-IUU regulations, the EU may dilute the strong fisheries focus of IUU. Moreover, putting human and labour rights into the anti-IUU trade measures, the EU risks the move 'backfiring' in that the trade measures may then come to violate WTO principles (Leroy et al., 2016; Wongrak et al., 2021). To our knowledge, the Thailand case remains the only unorthodox instance where labour rights have been included as part of fisheries dialogue (Kadfak & Linke, 2021).

While this next point has already been taken up in the chapter on Thailand, it is important to emphasize here again, how the policy diffusion lens allows us to unpack the labour 'add-on' during the EU-Thai dialogue for fisheries reform. In this case, the external policy of EU anti-IUU regulation has become a reflexive policy version of the EU's internal political agenda on human and labour rights within seafood trade policy. The non-linear nature of policy spread on labour standards is not new. The EU has been working with ILO labour standards and the ILO as an active agency to promote the 'European Social Model' where labour standards should be advanced through external activities and trade policy (Orbie, 2011).

Instead of pushing for its own definition of IUU fishing, the EU has used well-accepted descriptions of IUU from international organisations, in ways that benefit the EU. According to Gilardi and Wasserfallen (2019), it is important to explore the politics of policy diffusion in the early stages of the policy cycle. This is the stage for 'issue definition', where policy creators can change the terms of political debate, creating taboos or increasing the acceptance of ideas in mainstream political discourse (ibid, p. 1250). The EU anti-IUU policy is often known as the 'heavy-weight' of anti-IUU policies, following the United Nations Food and Agriculture Organisaion (FAO) International

Plan of Action to Prevent, Deter and Eliminate Illegal, Unreported and Unregulated Fishing (IPOA-IUU) and the Code of Conduct for Responsible Fisheries. In other words, the EU's anti-IUU regulation replicates international norms and guidelines, which were established before the birth of EU anti-IUU policy and have generally been well accepted in international affairs. In this way the EU created an airtight connection between its anti-IUU policy and the 'issue definition' on IUU in existing international discourses on the wicked problem of IUU fishing globally. This point has come out quite clearly from our interviews with the EU officers that Thailand should comply with existing measures on the conservation and management of Regional Fishery Management Organisations (RFMOs) and, where relevant, international laws that address IUU fishing, and there is no specific requirement to follow particular EU regulations.

Thailand domestic politics

Policy diffusion has never existed in a policy vacuum in the recipient country. The existing national-level policy field has a major influence on how diffused policy ideology and practices become materialised and institutionalised (Song et al., 2019). Thailand's political situation during the time of receiving the yellow card – with a military junta government being in power – shaped the rapid and corporatist response towards pressure from the EU for significant change in fisheries governance, including human and labour rights in the seafood sector. The military government responded proactively to improve the image of the country as a seafood producer and processor, to prevent further trade sanctions in other sectors (Auethavornpipat, 2017). One of our informants (an EU representative) opined that the outcomes of the reform would have been different, much slower, if the Thai government at the time had come from a democratic election. Electoral party politics would potentially have prevented the proactive determination to 'get rid' of the yellow card that was demonstrated by the military government. The case of Thai reforms shows how domestic processes and internal factors can facilitate, block and otherwise influence the trajectory of policy diffusion (Song et al., 2019, p. 139).

The Australia case

Australia's position in relation to anti-IUU fishing trade measures is different to Thailand's in that Australia is not a significant exporter to the EU or the US, and thus has not been the focus of anti-IUU import rules as a coercive measure. Australia is not in competition with Thailand and other countries exporting to the EU or US, so does not need to 'keep up' with the compliance of other exporting countries, nor is Australia in direct competition with the EU, US or Japan as large seafood importing markets. Economic competition is thus also not a potential driver for Australia to adopt anti-IUU fisheries trade restrictions. In this book we have considered Australia as a jurisdiction that might consider implementing similar kinds of trade restrictions as have been applied by the EU, US and Japan. In this way Australia is a potential 'receiver' of policy diffusion by the EU, US, and Japan, through learning or emulation mechanisms of diffusion.

Relative to other countries discussed in this book Australia is a small seafood trading country and has small per capita consumption of seafood (Department of Agriculture, Fisheries and Forestry, 2022). The adoption of anti-IUU trade rules by Australia would thus not have a major impact on global seafood trade flows. The seafood supply chains in which Australia is implicated, however, are very relevant for the question of using anti-IUU trade restrictions. Australia's seafood imports mainly come from Thailand, China, Vietnam and New Zealand, and its exports mainly go to China, Vietnam, Japan and Hong Kong (Steven, Mobsby & Curtotti, 2020). Thailand, China (including Hong Kong) and Vietnam are all countries for which IUU concerns have been raised internationally – with Thailand having been through an EU carding process and Vietnam facing this challenge at the time of writing. The EU has chosen not to apply its anti-IUU measures to imports of seafood from China, but China has a poor reputation regarding IUU, scoring the highest of any country in the world on the global IUU index (Macfadyen & Hosch, 2021). If Australia were to adopt similar kinds of anti-IUU trade restrictions as the EU, US or Japan, it could greatly change seafood markets within Australia. Moreover, Australia was active in the international sphere in creating the norm of anti-IUU and in developing measures to combat IUU, including import restrictions, and so seemed a likely candidate for adopting the policy itself, but it did not. The Australia case is thus instructive for this book because it provides material for considering the limitations of anti-IUU trade rules diffusing broadly to other countries.

Australia and anti-IUU fisheries trade measures

The Australian government was an active party driving international initiatives against IUU through the United Nations Food and Agriculture Organisation (FAO), such as the Port States Measures Agreement (PSMA) and the International Plan of Action against Illegal, Unreported and Unregulated fisheries (IPOA-IUU), both of which foreground the prevention of IUU catch from reaching markets. For example, in 2000 the Government of Australia hosted, with the FAO, an expert consultation on IUU that was foundational to the International Plan of Action (FAO, 2001). Australia was one of the early countries to sign and ratify the PSMA (FAO, 2023). Australia was also active in creating catch documentation schemes for Patagonian toothfish under the Convention for the Conservation of Antarctic Marine Living Resources (CCAMLR), and for southern bluefin tuna under the Convention for the Conservation of Southern Bluefin Tuna (CCSBT). These catch documentation schemes were specifically to allow importing countries to refuse to import undocumented toothfish or southern bluefin tuna.

The schemes have arguably been very effective. The CCAMLR catch documentation scheme used to prevent IUU fish being imported is widely accepted as having been one of the factors in successfully reducing the unsustainable levels of fishing on Patagonian toothfish since its adoption in 1999 (CCAMLR, 2023). Under the CCSBT management regime Southern bluefin tuna stocks have started to recover from their badly overfished state, with the IUCN redesignating the species from Critically Endangered to Endangered in 2021 (IUCN, 2021). After having been so active in developing anti-IUU policies, and in the face of evidence that importation restrictions could be effective in reducing IUU, the fact that Australia did not then adopt anti-IUU trade policy itself raises questions about the limitations of policy diffusion.

The emulation mechanism of policy diffusion, Australia and anti-IUU fisheries trade measures

The policy diffusion literature has tended to assume that adoption is based on rational calculations of whether a policy is effective in achieving objectives. Gilardi and Wasserfallen (2019, p. 1278) point out that most studies of diffusion assume that policy makers adopt policies because they learn that the policy is effective, or they feel it is necessary adopt a policy to avoid unpleasant consequences from a coercive process or through losing out in competition with other states. Emulation has not been written about as much as the other three types of diffusion. Emulation has been the category for motivations for policy adoption based the perceived morality or appropriateness of policies, rather than on evidence-based assessments of the success or failure of policies applied elsewhere (Gilardi & Wasserfallen, 2019, p. 1249).

The conceptualisation of emulation as a policy diffusion mechanism has been influenced by social constructivist thinking, which has focussed on the role of norms and conventions in policy spheres, such as international agencies and organisations, inspiring policy makers to adopt policies that conform with these norms (Finnemore & Sikkink, 2001). The spread of human rights policies following the Universal Declaration of Human Rights (1948) is a prominent example of a policy diffusing because it is seen as the 'right thing to do'. Anti-IUU policy fits within this conceptualisation of policy diffusion by emulation, with norms against IUU having been generated and promoted through deliberations of the FAO, leading to conventions such as the PSMA, IPOA-IUU and through CCAMLR and CCSBT and their management measures. IUU was named as a problem and the fight against IUU raised as an important fisheries management norm within these international organisations, developing further into measures including trade restrictions, which then went on to be adopted in member states.

From a constructivist perspective, Australia seemed well placed to emulate the EU anti-IUU seafood import policy. The Australian government actions noted above and the quotes presented in Chapter 3 clearly show that Australian fisheries policy-makers shared the norm that IUU should be tackled with various tools, including trade restrictive measures. At the time of writing, however, the emulation mechanism had not been strong enough to cause Australia to adopt the policy. The Australia case therefore shows one kind of limitation to policy diffusion by emulation – policy actors may promote a policy as appropriate internationally, and for other states, but see it as not being appropriate for themselves. Thinking through why Australia did not adopt anti-IUU trade measures reveals more about the limitations to policy diffusion in this case.

Why did Australia not adopt anti-IUU fisheries trade measures?

The materials we have examined for the Australian case in this book do not provide a clear answer on why Australia did not adopt anti-IUU fisheries trade measures, when it promoted them internationally. Australian government statements on the topic reveal some policy incoherence. For example, the federal Department of Agriculture, Fisheries and Forestry (DAFF) said in its submission to the Inquiry on Country of Origin Labelling (Commonwealth of Australia, 2014) that DAFF opposed introducing any sustainability measures on imports, for reasons including that they could be trade restrictive, and that they increased regulatory burden, noting that Australian exporters were suffering from the regulatory burden of import measures of other countries (Submission 11).[2] DAFF is responsible for Australia's anti-IUU actions internationally and so had been the agency active in developing trade measures for toothfish and southern bluefin tuna, in part to promote the interests of Australian fishing companies involved in these fisheries. Opposing Australia's adoption of the measure domestically is thus inconsistent with DAFF's actions internationally.

It is possible that ideological politics may play a role in this case of a failure of policy diffusion, with the Labor Party that came to power in 2022 more willing than the previous conservative Liberal National coalition government to consider regulation on imports. The material to hand at the time of writing, however, is not sufficient to make a case either way regarding ideological politics and policy diffusion of anti-IUU fisheries trade measures. Some of the explanation of why Australia refused to adopt anti-IUU policy seems to lie in the relative capacities of different interest groups to secure policy support. The politics of influence varies among large-scale, export-oriented fisheries, small-scale fisheries targeting domestic markets, seafood importing businesses and the hospitality sector. Another part of the explanation seems to lie in the domestic administrative and jurisdictional structure not being amenable to applying fisheries measures further along the supply chain, and institutional inertia against change.

Interest group politics

In addition to combatting IUU, another prominent objective for EU and US measures to prevent IUU products entering domestic markets is to 'level the playing field' between domestically produced seafood subject to regulation to prevent overfishing, and imported seafood which may not have been subject to the same level of regulation (Damanaki & Lubchenco, 2012). The anti-IUU trade measures in the EU and US, then, constitute government support for their respective fishing industries. By contrast, the Australian National Plan of Action on IUU (Department of Agriculture, 2014) is silent on using anti-IUU measures to level the playing field between domestically produced seafood and imports. Australian fishing industry groups as well as non-government organisations have argued that it is likely that some of the seafood imported to Australia is from IUU fisheries and is disadvantaging regulated domestic producers, giving evidence from shark fisheries.[3]

Australia's split position on anti-IUU trade measures – supporting them for overseas markets but not adopting one itself – in effect gives differing levels of support to different segments of the Australian fishing industry. In promoting anti-IUU measures for Patagonian toothfish and Southern bluefin tuna, the Australian government supported the interests of large-scale, export-oriented Australian companies that were suffering from international IUU fishing in these fisheries. But in refusing to adopt anti-IUU trade measures itself, the Australian government has declined to support the smaller-scale segment of the Australian fishing industry that sells in domestic markets and competes against cheaper imported seafood.

The Australian government's choice not to support with anti-IUU trade measures the smaller-scale, less profitable segment of the fishing industry that supplies domestic markets aligns with various other fisheries management policy choices at state and federal levels since the 1990s. Policies have tended to shrink the numbers of commercial fishing operators, making unviable the smaller, less profitable, diversified (by gear and target species) operations and favouring larger, specialist, more profitable fishing operators. Incentivising certain types of fishing business and discouraging other business models has occurred through the application of individual transferable quotas and other management measures that require companies to undertake sophisticated administrative reporting (Fabinyi & Barclay 2022; Minnegal & Dwyer 2008). As an interest group, the segment of the fishing industry selling in domestic markets that would benefit from anti-IUU fishing regulation on imports and from Country of Origin Labelling (CoOL), seems to have less weight with the Australian government than seafood importers and the hospitality sector, who benefit from cheap imports. Several government agencies opposed introducing CoOL for seafood in food service industries on the grounds that it would increase the regulatory burden for seafood importers and retailers (Submissions 4, 19, 11).

The Australian case thus shows that policy diffusion is mediated by the politics of influence between interest groups who would be differently affected by the policy with the receiving government. Further, the Australian case shows that this politics of influence can be complex. The Australian government has not supported the seafood industry as a whole, or even the fishing industry as a whole, but has acted in a way that supports the interests of some parts of the industry and disadvantages others.

Jurisdictional boundaries

Policy diffusion is also related to domestic regulatory frameworks. Arguably Australia's domestic regulatory framework was not very amenable to regulation for fisheries using trade measures, because fisheries regulation had hitherto been restricted to the harvest node of the supply chain. Once seafood leaves the harvest node it is regulated by government mainly in terms of food safety. The legality of catch is largely not regulated in markets. The related issue of the sustainability of the mode of fishing is portrayed by government actors as something that should not be regulated by government but be left to consumer choice (Garcia Garcia, Barclay & Nicholls 2020).

This means that when thinking about the receiving government and the potentials for policy diffusion, we must disaggregate the state. Different domestic agencies have varied responsibilities, roles, interests and priorities. The health agencies that have thus far been the ones responsible for regulating seafood in domestic markets are not interested in the sustainability or legality of fish harvesting, nor do they have the capacity to regulate for it. Submissions made by Australian federal and state fisheries management agencies to the CoOL Inquiry reiterate the position that fisheries management in Australia occurs in the fishing node of the supply chain, not at the importing or market end of the chain, and they showed no willingness to change this situation (Submissions 4, 11). This is quite different to the EU and US, where the anti-IUU importation regulations are under the aegis of fisheries management – they are administered in the EU by the Directorate-General for Maritime and Fisheries (DG Mare) and the US SIMP was legislated under the *Magnuson-Stevens Fishery Conservation and Management Act.* A substantive change in Australian fisheries management and seafood regulatory bodies' perceptions of their respective responsibilities would be needed for anti-IUU seafood importation regulations to be developed and implemented.

The possibility of institutional flexibility is visible in the changing uses of traceability mechanisms in the EU, where tools for food safety have been co-opted for use in anti-IUU. The legality of fish catches for EU and US importation purposes is now traced via catch documentation schemes (He, 2018, Helyar et al., 2014). The EU anti-IUU traceability system was developed from the traceability regulations already in place for food safety (Lewis & Boyle, 2017). The Australian National Plan of Action on IUU (Department of Agriculture, 2014), however, precluded the use of traceability for anti-IUU efforts, limiting its use to food safety purposes. Again, the demarcation between different government roles and purposes was used as a reason for Australia to not adopt the policy.

***

In sum, we can say several things about the limitations to policy diffusion revealed by the Australian case. The conditions were conducive for anti-IUU fisheries trade policy to diffuse to Australia, by either emulation or learning mechanisms, but the policy was not adopted. The reasons the policy did not diffuse included political reasons, such as the interest group that could have benefited from anti-IUU trade policy being relatively less influential than other interest groups that did not want the policy to be established. Moreover, the Australian 'state' for the purposes of policy diffusion was not a unitary actor, but was made up of agencies. Jurisdictional boundaries that prevented policy adoption were adhered to and agencies refused to change. The situation may change with the change of government in 2022, from a conservative to a more liberal party. The new Labor Party Minister for Fisheries has opened up discussion on the idea of anti-IUU fishing import regulations.

Conclusion

Our book is one of the first focusing on the 'reception' of EU anti-IUU policy in other countries, using the conceptual framework of policy diffusion to examine the adoption potential of the EU trade restrictive regulations as a means of curtailing the occurrence of IUU fishing globally. Based on primary empirical data, the book performs an analysis of how two countries – Thailand and Australia – have dealt with such a measure, and examines what kind of domestic processes have driven outcomes and their respective outlooks. The two case studies we present in this book reflect how the global community concerned with policy settings can expect future implementation of the trade-based regulation in other countries to control IUU fishing to unfold. From these analyses, our book offers answers on how countries will adapt to changing policy norms regarding IUU fishing.

We have elaborated concrete examples of how two countries, positioned differently on the receiving end of the policy, engage with EU implementation. Understanding the (re)actions of "other" countries, who might be targeted or otherwise influenced by the policy, will form an essential new knowledge that helps inform a more effective and scalable implementation of the policy on the part of the EU, and a better governance preparedness on the part of non-EU fishing nations. In particular this book exposes a key moment of change in the compatibility between environmental regulations and international trade. The EU anti-IUU policy is a prime example of a policy that uses the mechanisms of international trade to account for environmental and conservation objectives. By way of the unilateral and trade-restrictive stance against IUU fishing, the EU has positioned itself as a major market and normative power, driving its sustainability norms outwards. This book sheds light on the efficacy of this policy setup based on the analysis of country perspectives, which is a key factor influencing its potential spread.

While the main focus of this book is the potential for the policy to spread and the impact of trade measures in the two cases, it is also important to reflect further on what has been revealed by our in-depth analysis of each case. We observe that the outcomes of implementing anti-IUU trade measures is not restricted to trade pressures and the promotion of anti-IUU ideology in third countries and global politics. Rather, the EU anti-IUU policy has also exerted normative and ideological powers that have been influential in Australia's and Thailand's domestic policy regarding IUU fishing. This book therefore takes a step further than Sumaila's (2019) study on the economic risks of IUU fishing in the context of the EU carding system. We agree with Sumaila's work that major seafood markets like US and Japan should adopt anti-IUU trade measures to enable the regime to have a global impact. However, our study also shows that the Thai governments worked hard to avoid the reputational risk of being seen as a 'bad actor' in IUU fishing globally. Economic risk, therefore is part of a bigger picture in which less tangible forms of power over reputation are also at play.

Moreover, exploring Australia as a potential adopter of the EU IUU policy is a novel approach, since we might expect more middle-ranking importing countries would feel pressure to act in line with major importing countries and the new fisheries norm to address IUU problems with market power. Without a cohesive global policy on trade from IUU fisheries, seafood from IUU fisheries can still find markets. Kadfak learned from fieldwork to Thailand in early 2023 that some seafood caught in Thailand has deviated away from the EU and US to other markets in Asia, including China, which have less restrictive requirements. Researchers have started exploring how China as a major seafood market and fishing nation could shape IUU fishing, albeit through incoherent policy (Song, Fabinyi & Barclay, 2022). So far, we know very little about how the EU anti-IUU policy has affected South-South seafood trade flows to avoid strict regulation. This aspect of implications from the policy requires further study. Moreover, a more long-term approach to explore policy adoption via bilateral dialogues is needed to truly understand the impact of EU anti-IUU policy beyond EU borders.

Finally, this book provides insights as to how the governance interactions between the EU and other fish exporting/importing nations might need to be adjusted to improve the effectiveness of the aforementioned policy. The book offers thoughts on whether the current mode of implementation provides a scalable solution towards reducing global incidences of IUU fishing. We hope that this book will provide a unique perspective on IUU fishing from two different receiving ends, in ways that illustrate how and with what consequences a unilateral environmental policy aimed at discouraging IUU fishing actually plays out in other countries which might be expected to align or comply with the EU policy direction.

Notes

1 For more information about the EU IUU fishing Coalition, please see https://www.iuuwatch.eu/about/ (retrieved February 5, 2023).
2 The submission documents are available at https://www.aph.gov.au/Parliamentary_Business/Committees/Senate/Rural_and_Regional_Affairs_and_Transport/Seafood_labelling/Submissions (retrieved February 1, 2023).
3 The argument about imports from IUU fisheries disadvantaging regulated Australian shark fisheries was made by the Southern Shark Industry Alliance (SSIA) and Traffic International (Submission 13) and the seafood industry

Common Language Group, under the Fisheries Research and Development Corporation (Submission 17 Attachment 2) in submissions to the CoOL Inquiry. See: https://www.aph.gov.au/Parliamentary_Business/Committees/Senate/ Rural_and_Regional_Affairs_and_Transport/Seafood_labelling/Submissions (retrieved February 1, 2023).

References

Auethavornpipat, R. (2017). Assessing regional cooperation: ASEAN states, migrant worker rights and norm socialization in Southeast Asia. *Global Change, Peace & Security, 29*(2), 129–143.

Bhakoo, V., & Meshram, K. (2021). Modern slavery in supply chains. In *The Routledge Companion to Corporate Social Responsibility* (pp. 268–280). Abingdon UK: Routledge.

Braun, D., & Gilardi, F. (2006). Taking 'Galton's problem' seriously: Towards a theory of policy diffusion. *Journal of theoretical politics, 18*(3), 298–322.

CCAMLR (2023). Elimination of IUU fishing and the World's First Catch Documentation Scheme [web page]. Commission for the Conservation of Antarctic Marine Living Resources (CCAMLR). Retrieved March 6, 2023 from, https://40years.ccamlr.org/elimination-of-iuu-fishing-and-the-worlds-first-catch-document-scheme/#:~:text=In%201999%2C%20CCAMLR%20developed%20the,Documentation%20Scheme%20for%20Dissostichus%20spp

Commonwealth of Australia (2014). *Current requirements for labelling of seafood and seafood products*. Canberra: Parliament of Australia. Retrieved March 5, 2023 from, https://www.aph.gov.au/Parliamentary_Business/Committees/ Senate/Rural_and_Regional_Affairs_and_Transport/Seafood_labelling

Damanaki, M. & Lubchenco, J. (2012). *Joint statement by European Union Commissioner for Maritime Affairs and Fisheries Maria Damanaki and United States Under Secretary of Commerce for Oceans and Atmosphere Jane Lubchenco, PhD.*, press release, Brussels: European Commission. Retrieved December 12, 2019 from, http://europa.eu/rapid/press-release_MEMO-12-382_en.htm?locale=en

Department of Agriculture (2014). *Australia's second national plan of action to prevent, deter and eliminate illegal, unreported and unregulated fishing*. Canberra: Department of Agriculture. Retrieved February 27, 2023 from, https://www.agriculture.gov.au/sites/default/files/sitecollectiondocuments/fisheries/iuu/aus-second-npoa-iuu-fishing.pdf

Department of Agriculture, Fisheries and Forestry (2022). Fisheries and aquaculture statistics website, ABARES, Canberra. Retrieved January 31, 2023 from, https://www.agriculture.gov.au/abares/research-topics/fisheries/fisheries-and-aquaculture-statistics

EJF (2022). Driving Improvements in Fisheries Governance Globally: Impact of the EU IUU Carding Scheme on Belize, Guinea, Solomon Islands and Thailand. Environmental Justice Foundation (EJF). Retrieved March 13, 2023 from, https://ejfoundation.org/resources/downloads/EU-IUU-Coalition-Carding-Study.pdf

Fabinyi, M., & Barclay, K. (2022). *Asia-Pacific Fishing Livelihoods*. Palgrave Macmillan.

FAO (2001). International Plan of Action to Prevent, Deter and Eliminate Illegal, Unreported and Unregulated Fishing. Food and Agriculture Organisation of the United Nations (FAO), Rome. Retrieved March 6, 2023 from, https://www.fao.org/3/y1224e/y1224e.pdf

FAO (2023). FAO Treaties Database, Agreement on Port State Measures to Prevent, Deter and Eliminate Illegal, Unreported and Unregulated Fishing (PSMA) [website]. Retrieved March 6, 2023 from, https://www.fao.org/treaties/results/details/en/c/TRE-000003/

Finnemore, M., & Sikkink, K. (2001). Taking Stock: The Constructivist Research Program in International Relations and Comparative Politics. *Annual Reviews in Political Science, 4*, 391–416.

Garcia Garcia, S., Barclay, K., & Nicholls, R. (2020). The multiple meanings of fish: Policy disconnections in Australian seafood governance. In E. Probyn, K. Johnston, & N. Lees (Eds.), *Sustaining seas: Oceanic space and the politics of care* (pp. 75–86). London & New York: Rowman & Littlefield.

Gilardi, F., & Wasserfallen, F. (2019). The politics of policy diffusion. *European Journal of Political Research, 58*(4), 1245–1256.

He, J. (2018). From country-of-origin labelling (COOL) to seafood import monitoring program (SIMP): How far can seafood traceability rules go? *Marine Policy, 96*, 163–174. 10.1016/j.marpol.2018.08.003

Helyar, S. J., Lloyd, H. A. D., de Bruyn, M., Leake, J., Bennett, N., & Carvalho, G. R. (2014). Fish product mislabelling: Failings of traceability in the production chain and implications for Illegal, Unreported and Unregulated (IUU) fishing. *PLoS ONE, 9*(6). 10.1371/journal.pone.0098691

IUCN (2021). Tuna species recovering despite growing pressure on marine life – IUCN Red List. Press release, September 4. International Union for the Conservation of Nature (IUCN). Retrieved March 6, 2023 from https://www.iucn.org/news/species/202109/tuna-species-recovering-despite-growing-pressures-marine-life-iucn-red-list

Johnson, B., & Hagström, B. (2005). The translation perspective as an alternative to the policy diffusion paradigm: The case of the Swedish methadone maintenance treatment. *Journal of Social Policy, 34*(3), 365–388. doi: 10.1017/S0047279405008822

Kadfak, A., & Antonova, A. (2021). Sustainable Networks: Modes of governance in the EU's external fisheries policy relations under the IUU Regulation in Thailand and the SFPA with Senegal. *Marine Policy, 132*. 10.1016/j.marpol.2021.104656

Kadfak, A., & Linke, S. (2021). Labour implications of the EU's illegal, unreported and unregulated (IUU) fishing policy in Thailand. *Marine Policy, 127*. 10.1016/j.marpol.2021.104445

Kadfak, A., Wilhelm, M., & Oskarsson, P. (2023). Thai Labour NGOs during the 'Modern Slavery' Reforms: NGO Transitions in a Post-aid World. *Development and Change.* 10.1111/dech.12761

Leroy, A., Galletti, F., & Chaboud, C. (2016). The EU restrictive trade measures against IUU fishing. *Marine Policy, 64*, 82–90. 10.1016/j.marpol.2015.10.013

Lewis, S. G., & Boyle, M. (2017). The expanding role of traceability in seafood: Tools and key initiatives. *Journal of Food Science, 82*, A13–A21. 10.1111/1750-3841.13743

Macfadyen, G. & Hosch, G. (2021). The IUU Fishing Index, 2021. Poseidon Aquatic Resource Management Limited and the Global Initiative Against Transnational Organized Crime. Retrieved March 6, 2023 from, https:// iuufishingindex.net/profile/china

Manners, I. (2002). Normative power Europe: a contradiction in terms? *JCMS: Journal of common market studies, 40*(2), 235–258.

Manners, I. (2011). The European Union's normative power: critical perspectives and perspectives on the critical. In Whitman, R. G. (ed.), *Normative Power Europe: Empirical and Theoretical Perspectives* (pp. 226–247). Basingstoke UK: Palgrave MacMillan.

Miller, A. M., Bush, S. R., & Mol, A. P. (2014). Power Europe: EU and the illegal, unreported and unregulated tuna fisheries regulation in the West and Central Pacific Ocean. *Marine Policy, 45*, 138–145. 10.1016/j.marpol.2013.12.009

Minnegal, M., & Dwyer, P. D. (2008). Managing risk, resisting management: stability and diversity in a Southern Australian fishing fleet. *Human Organization,* 67(1), 97–108. doi:10.17730/humo.67.1.x38g60k463p26855.

Mukhtarov, F. (2014). Rethinking the travel of ideas: Policy translation in the water sector. *Policy & Politics, 42*(1), 71–88. 10.1332/030557312X655459

Mundy, V. (2018). *The impact of the EU IUU Regulation on seafood trade flows: Identification of intra-EU shifts in import trends related to the catch certification scheme and third country carding process.* Brussels, Belgium: Environmental Justice Foundation, Oceana, The Pew Charitable Trusts, WWF. Retrieved March 23, 2023 from, https://europe.oceana.org/wp-content/uploads/sites/26/ tda_report_iuuwatch_hq.pdf

Orbie, J. (2011). Promoting labour standards through trade: normative power or regulatory state Europe? In Whitman, R. G. (ed.) *Normative Power Europe: Empirical and Theoretical Perspectives* (pp. 161–184). Basingstoke UK: Palgrave MacMillan.

Peck, J. (2011). Geographies of policy: From transfer-diffusion to mobility-mutation. *Progress in Human Geography, 35*(6), 773–797. 10.1177/0309132510394010

Steven, A. H., Mobsby, D., & Curtotti, R. (2020). *Australian fisheries and aquaculture statistics 2018* (Fisheries Research and Development Corporation project 2019-093), Canberra: ABARES. Retrieved February 27, 2023 from, https://www.frdc.com.au/project/2019-093

Shipan, C. R., & Volden, C. (2008). The mechanisms of policy diffusion. *American journal of political science, 52*(4), 840–857.

Simmons, B. A., Dobbin, F., & Garrett, G. (2006). Introduction: The international diffusion of liberalism. *International organization, 60*(4), 781–810.

Song, A. M., Cohen, P. J., Hanich, Q., Morrison, T. H., & Andrew, N. (2019). Multi-scale policy diffusion and translation in Pacific Island coastal fisheries. *Ocean & Coastal Management, 168*, 139–149. 10.1016/j.ocecoaman.2018.11.005

Song, A. Y., Fabinyi, M., & Barclay, K. (2022). China's approach to global fisheries: power in the governance of anti-illegal, unreported and unregulated fishing. *Environmental Politics,* 1–21. 10.1080/09644016.2022.2087338

Sumaila, U. R. (2019). A carding system as an approach to increasing the economic risk of engaging in IUU fishing? *Frontiers in Marine Science, 6.* 10.3389/ fmars.2019.00034

Thorpe, A., Hermansen, O., Pollard, I., Isaksen, J., Failler, P., & Touron-Gardic, G. (2022). Unpacking the tuna traceability mosaic–EU SFPAs and the tuna value chain. *Marine Policy*, *139*, 105037. 10.1016/j.marpol.2022.105037

Wilhelm, M., Kadfak, A., Bhakoo, V., & Skattang, K. (2020). Private governance of human and labor rights in seafood supply chains–The case of the modern slavery crisis in Thailand. *Marine Policy*, 115, 103833. 10.1016/j.marpol.2020. 103833

Wipatayotin, A. (2019). Prawit urges UN to get tough on illegal fishing. *Bangkok Post*. Retrieved February 21, 2023 from, https://www.bangkokpost.com/thailand/general/1689872/prawit-urges-un-to-get-tough-on-illegal-fishing.

Wongrak, G., Hur, N., Pyo, I., & Kim, J. (2021). The Impact of the EU IUU Regulation on the Sustainability of the Thai Fishing Industry. *Sustainability*, *13*(12), 6814. 10.3390/su13126814

Index

Printed in the United States
by Baker & Taylor Publisher Services